ALSO BY PIERS PAUL READ

NONFICTION

Alive: The Story of the Andes Survivors
The Train Robbers

FICTION

Game in Heaven with Tussy Marx
The Junkers
Monk Dawson
The Professor's Daughter
The Upstart
Polonaise
A Married Man
The Villa Golitsyn
The Free Frenchman
A Season in the West
On the Third Day

ABLAZE

THE STORY OF

Random House ⌂ New York

ABLAZE
THE HEROES AND VICTIMS
OF CHERNOBYL

Piers Paul Read

Library of Congress Cataloging-in-Publication Data
Read, Piers Paul
Ablaze: The story of the heroes and victims of Chernobyl/by Piers
Paul Read
    p.   cm.
ISBN 0-679-40819-3
1. Chernobyl Nuclear Accident, Chernobyl, Ukraine, 1986—
Environmental aspects.   2. Radioactive pollution—Environmental
aspects—Ukraine—Pripyat.   3. Radiation injuries—Ukraine—
Pripyat.   I. Title.
TD196.R3R42   1993   363.17′99′0947714—dc20   92-56840

Manufactured in the United States of America
98765432
First Edition

Book Design by Oksana Kushnir

But oh! It matters much to me
If wicked men with cunning ways
Should lull our fair Ukraine to sleep,
To better plunder, set ablaze. . . .
Oh yes, it matters much to me.

Taras Shevchenko
Haidamaki (1841)

ACKNOWLEDGMENTS

In writing this book I am indebted first and foremost to my Russian research assistant, Natasha Segal, who assembled all the published sources from different archives in Britain, Germany and the Soviet Union and arranged the two research trips that I made to Russia, Belorussia and the Ukraine in the spring and autumn of 1991. Despite the declassification of information about the Soviet atomic industry two years before, it required exceptional tact and skill first to find the different witnesses, and then to persuade them to talk to a Western writer. Although certain liberties had been established by the time we started our research in April 1991, there remained a real fear of a return to the former regime, and also a reluctance among some to reveal the shortcomings of their country.

I am also grateful to Robert S. Tilles, president of the relief organization Chernobyl Help, in Moscow, who extended the necessary invitations; to Volodomyr Shovkovshytny, president of the Chernobyl Union in Kiev; to his tireless assistant, Lyudmila Golybardova, and her husband, Leonid; to Lubov Kovalevskaya; to Nikolai Steinberg, then deputy chairman of the USSR State Committee for the Supervision of Safety in the Nuclear Power Industry; and Professor Anatoli G. Nazarov, chairman of the Supreme Soviet Committee of Investigation into the Causes of the Chernobyl Catastrophe.

I should also like to record my gratitude to all those in the former Soviet Union who spared the time to talk to me and who frequently, despite the scarcities of food and drink at the time, extended a hospitality that did honor to their tradition.

In alphabetical order, these were: Dr. Armen A. Abagyan, Academician Anatoli Alexandrov, Ivan and Irina Avramenko; Vladimir Babichev, Professor Alexander Baranov, Drs. Anatoli and Tatiana Ben, General Gennadi Berdov,

Judge Raimond Brize, Victor and Valentina Brukhanov, Boris Chaivanov, Nina Chernaieva, W. Choleska, Vladimir and Nina Chugunov, Alexei and Antonia Dashuk, Razim and Inze Davletbayev, Academician Nikolai Dollezhal, Professor Boris Dubowski, Anatoli and Isabella Dyatlov, Konstantin Fedulenko, Vladimir Fedulov, Professor Andrei Gagarinski, Vadim and Ylena Grishenka, Professor Dmitri Grodzinski, Vladimir Gubarev, Alexander Gulevsky, Dr. Angelina Guskova, Dr. Ylena Holod, Academician Leonid Ilyn, Professor Yuri Israel, Professor Eugene Ivanov, Alexander Kalugin, Mykola Karpan, Anna Karpushova, Dr. Adolf Kharash, Ivan Khodeley, Igor Kirschenbaum, Vladimir Klimov, Professor Victor Knijnikov, Eduard Korotkov, Igor Kravchenko, Professor Yuri Ladzurkin, Lubov Lelelchenko, Vasily Lesovoy, Tania Likhacheva, Katya Litovsky, Vladimir Lukin, Dr. Vladimir Lupandin, Nadezda Lutchenko, Professor Alexander Lutzko, Alexander Nemirovsky, Tatyana Nevmergikaya, Anatoli Nevmerzhitsky, Georgi Nogaevsky, Anna Odinokaya, Piotr and Tatiana Palamarchuk, General Vladimir Pikalov, Professor Nikolai Protzenko, Dr. Anatoli Romanenko, Professor Eugene Ryzantzev, Nikita Schlovsky, Tatiana Seleviorsteva, Yuri Shadrin, Dimitry Shatalov, Ivan Shavrey, Leonid Shavrey, Vladimir Siroshtan, Professor Yuri Sivintsev, Efim Slavsky, Mikhail Stelmak, Dr. Evgenia Stepanovna, Professor Arcadi J. Svirnovski, Fyodor and Olga Tithonenko, Arkadi Uskov, Academician Yevgeni Velikhov, Professor Andrei Vorobyov, Alexander and Natasha Yuvchenko.

In Kiev, I was given useful advice by Michael Behr of the International Red Cross, Cornelia Wendt of the Bavarian Red Cross, and Dr. Edmund Lengfelder of the Radiobiological Institute of Munich University. Alexander Vlasenko kindly acted as interpreter on my visit to Chernobyl. I am also grateful to those who helped me in Moscow, particularly Serge and Manya Schmemann and Olga Voronovna. I would also like to thank Professor Richard Wilson and Marina Kostanetskaya, who gave interviews to Natasha Segal.

In Britain, I was given invaluable assistance on the technical aspects of the story by Lorna Arnold, John Collier, Ken Duncan and Bryan Edmondson, who read my manuscript and gave me the benefit of their considerable knowledge of, and long experience in, the nuclear field. They should not be held responsible for any errors that remain in the text or for any judgments that are implicit or explicit in what I have written. I would like to thank Robin Jones, Liza Dimbleby and Helena Fraser, who translated documents from Russian into English, and Michael Sarni, who made translations from Ukrainian. I am also grateful, for their advice, to David Morton, Dr. Anthony Daniels, Norma Percy, Paul Mitchell, Andro Linklater, Dr. Keith James, Marco Bojcun, Professor Terence Lee, Dr. John Marks, Constantine Zelenko, Gemma Hunter, and John Roberts.

I gratefully acknowledge permission to quote from *Chernobyl: A Documentary Story* by Iurii Shcherbak, translated from the Ukrainian by Ian Press, published by Macmillan in association with the Canadian Institute of Ukrainian Studies, University of Alberta; and from *The International Chernobyl Project: Proceedings of an International Conference* published by the International Atomic Energy Agency, Vienna.

I would like to thank the following for permission to use photographs: the Kurchatov Institute; Nikolai Dollezhal; Valentin Obodzinski; Vadim Grishenka; Nikolai Steinberg; Dmitri Shatalov; Angelina Guskova; Drs. Anatoli and Tatiana Ben; General Pikalov; Edward Korotkov; Leonid Ilyn; Lubov Kovalevskaya; Vladimir Chugunov; Dmitri Grodzinski; Raimond Brize; Yuri Shadrin; and Edward Korotkov.

Finally, I should like to thank Emily Read, my wife, for correcting the manuscript of this book; Bonnie Levy for her superb copyediting of the final draft; my editor at Random House, who suggested that I should write it; and Gillon Aitken, who, at certain difficult moments, persuaded me to persevere.

Contents

N

Belorussia

Novozybkov

Gomel

Mozyr

Hojniki   Bragin

Slavutich

Pripyat   Chernigov

Ovruch   **Chernobyl**

Polesskoe

Narodichi

Ivankov

Kiev

Zhitomir

Ukraine

30 km. zone of highest
contamination from
Chernobyl accident

Kms.
0   40
0   40
Miles

©1993 A. Karl / J. Kemp

Areas contaminated by radioactive cesium-137 released during the Chernobyl accident

N

SOVIET UNION

Leningrad

Moscow

RUSSIA

Smolensk

Tula

Orsa

Minsk

Mogilev

Brjansk

Orel

Belorussia

Gomel

Pinsk

Kursk

Chernigov

Chernobyl

UKRAINE

Kiev

Kharkov

Poltava

Miles
0   150

0   150
Kms.

© 1995 A. Karl / J. Kemp

THE FOUNDERS OF SOVIET ATOMIC POWER

Igor Kurchatov — Father of the Soviet atom bomb

Lavrenty Beria — Chief of the NKVD. Director of the atomic bomb project.

Efim Slavsky — Minister of Medium Machine Building

Nikolai Dollezhal — Designer of the first nuclear reactors. Director of NIKYET.

Anatoli Alexandrov — Director of the Kurchatov Institute. President of the Academy of Sciences.

Valeri Legasov — Academician. First deputy director of the Kurchatov Institute.

STAFF OF THE V. I. LENIN NUCLEAR POWER STATION AT CHERNOBYL

Victor Brukhanov — Director

Vasili Kizima — Head of construction

Nikolai Fomin — Chief engineer

Anatoli Dyatlov — Deputy chief engineer, 3rd and 4th units

Mikhail Lyutov — Scientific deputy chief engineer

The early morning shift on 26 April

Boris Rogozhkin — Chief of shift, 3rd and 4th units

Alexander Akimov — Shift foreman, 4th unit

Leonid Toptunov	Senior reactor control engineer
Piotr Stolyarchuk	Senior unit control engineer
Igor Kirschenbaum	Senior turbine control engineer
Yuri Tregub	Shift foreman prior to Akimov
Valeri Perevozchenko	Shift foreman, equipment maintenance department, 3rd and 4th units
Victor Proskuriakov	Engineer under Perevozchenko
Sasha Yuvchenko	Proskuriakov's replacement
Alexander Kudriatsev	Engineer
Gennady Metlenko	Engineer from Donenergo
Alexander Lelechenko	Head of the electrical workshop
Razim Davletbayev	Deputy head of the turbine section
Piotr Palamarchuk	Director of the start-up enterprise

The engineers from Komsomolsk

Anatoli Sitnikov	Deputy chief engineer, 1st and 2nd units
Vladimir Chugunov	Head of the reactor workshop, 1st and 2nd units
Vadim Grishenka	Deputy chief engineer, 5th block (under construction)

PRIPYAT

Vladimir Voloshko	Secretary of the city committee
A. S. Gamanyuk	Party secretary
Major Teliatnikov	Commander of the Pripyat and Chernobyl nuclear power station fire service
Lieutenant Pravik	Duty officer of the nuclear power station fire service on night of 25–26 April

Lieutenant Kibenok	Duty officer of the Pripyat fire service on the night of 25–26 April
Vitali Leonenko	Director of the Hospital
Anatoli Ben	Surgeon
Lubov Kovalevskaya	Acting editor of *Tribuna Energetica*

MOSCOW

The Central Committee

Mikhail Gorbachev	General secretary of the Communist Party of the Soviet Union
Nikolai Ryzhkov	Soviet prime minister (Chairman of the Council of Ministers of the USSR)
Yegor Ligachev	Secretary of the ideological department
Alexander Yakovlev	Secretary of the proaganda department
Vladimir Marin	Head of the nuclear power department

Defense and Civil Defense

Marshal Akhromeev	Chief of the Soviet general staff
General Altunin	Commander of the Soviet civil defense
General Ivanov	Second-in-command of the Soviet civil defense
Colonel General Pikalov	Commander of the chemical troops

Ministries and Institutes

Efim Slavsky	Minister of medium machine building
Anatoli Mayorets	Minister of energy and electrification
Armen Abagyan	Director of the All-Union Research Institute for Nuclear Power Plant Operation (VNIIAES)

Yuri Israel	Director of the State Committee of Hydrometeorology

The Kurchatov Institute

Anatoli Alexandrov	Director
Valeri Legasov	First deputy director
Yevgeni Velikhov	Deputy director. Expert on nuclear fusion.
Eugene Ryzantzev	Head of the nuclear safety department
Alexander Kalugin	Scientist responsible for RBMK reactors
Konstantin Fedulenko	Physicist
Yuri Sivintsev	Physicist

The Institute of Biophysics

Leonid Ilyn	Director
Victor Knijnikov	Head of the laboratory

In-patient Department of Institute of Biophysics at Hospital No. 6

Angelina Guskova	Director. Formerly Kurchatov's personal physician.
Alexander Baranov	Head of the hematology department
Georgi Seredovkin	Specialist in radiation sickness
Robert Gale	American bone-marrow transplant specialist

The Press

Vladimir Gubarev	Science editor of *Pravda*. Author of the play *Sarcophagus*.

The Chernobyl Commission

Boris Scherbina	Deputy prime minister. First chairman.
Ivan Silayev	Deputy prime minister. Second chairman.

Oleg Shepin — Deputy prime minister of health. Chairman of the medical commission.

Andrei Vorobyov — Hematologist. Principal expert on the medical commission.

KIEV

Officials

Vladimir Shcherbitsky — General secretary, Communist Party of the Ukraine

Grigori Revenko — First party secretary of the Kiev region

Vitali Sklerov — Ukrainian minister of power and electrification

Valentina Shevchenko — Ukrainian deputy prime minister

Anatoli Romanenko — Ukrainian minister of health

Defense and Civil Defense

General Berdov — Commander of the Ukrainian militia

Major General Antoshkin — Air force commander, Kiev region

Nationalists and Democrats

Yuri Shcherbak — Leader of the Green World movement in Kiev. People's deputy. Later minister of the environment.

Volodomyr Yavorivsky — Author. A people's deputy in the RUKH party. Framer of the Chernobyl Law.

Volodomyr Shovkovshytny — President of the Chernobyl Union. People's deputy in the Ukrainian Supreme Soviet.

Introduction

When Valeri Legasov, the scientist who took charge after the accident at Chernobyl, first saw the glowing crater of the ruined fourth reactor, he realized that he faced a disaster comparable to the San Francisco earthquake or the destruction of Pompeii.

Today, it seems likely that Chernobyl may be more significant than either of these historic catastrophes. Many millions have suffered and continue to suffer from the consequences of the accident. Parts of Russia, Belorussia and the Ukraine are uninhabitable and will remain so for hundreds, perhaps thousands of years to come. The count of the fatalities ranges from the official figure of thirty-one to a projection that Chernobyl will ultimately claim more victims than did World War II.

Equally significant is the effect of Chernobyl on our attitude toward nuclear power. Confidence was already shaken by the accident at Three Mile Island, after which no new nuclear power stations have been commissioned in the United States. In the wake of Chernobyl, some forty reactors in the Soviet Union that were either in operation, under construction or about to be built were closed, canceled or converted to thermal power. In Austria, a brand-new nuclear power station was mothballed, its cost paid by public subscription. In Italy, all plans for nuclear power were canceled. In Sweden, where the fallout from Chernobyl was first registered, the government made a commitment to phase out nuclear power by the year 2010.

However, an ever increasing demand for electricity, together with limited reserves of gas, coal and oil and the growing evidence of the damage done by their combustion to our health and the environment, makes this rejection of nuclear power itself problematic. For this reason, it is important to reach as full an understanding as possible of an event like the accident at Chernobyl.

Several books have already been written on Chernobyl by authors inside and outside the Soviet Union, but none that I know of by a Western writer since the fall of the Communists from power in 1991 and the subsequent opening of archives and loosening of tongues. Dr. Yuri Shcherbak, the author of *Chernobyl: A Documentary Story,* and Grigori Medvedev, the author of *The Truth About Chernobyl,* were both engaged in the aftermath of the accident—the first as a doctor, the second as an engineer—and so combine with their narrative the immediacy of a firsthand account. I have drawn on both to describe the role played by their authors. The same is true of *Final Warning: The Legacy of Chernobyl* by Dr. Robert Gale and Thomas Hauser.

Two expatriate Ukrainians, Viktor Haynes and Marko Bojcun, published *The Chernobyl Disaster* in 1988, an excellent factual account based on published sources. David Marples, from the Institute of Ukrainian Studies at the University of Alberta, published *Chernobyl and Nuclear Power in the USSR* in 1987 and has contributed regular well-informed articles on the subject to the *Report on the USSR.* More recently, there is *The Legacy of Chernobyl* by Zhores Medvedev, who once worked as a scientist in Soviet nuclear facilities and was expelled from his country in 1973. His book profits from his knowledge and experience and contains a more thorough explanation of the scientific aspects of the disaster than I have attempted here. Finally, *Chernobyl: Insight from the Inside* by Vladimir Chernousenko, a Ukrainian scientist who took a prominent part in dealing with the consequences of the accident, reflects the anxieties found in his country today.

For the political background that is so important in reaching a proper understanding of the story, I am indebted to *The Great Terror* by Robert Conquest; *Stalin: Triumph and Tragedy* by Dimitri Volkogonov; *Gorbachov: The Path to Power* by Christian Schmidt-Häuer; *The New Russians* by Hedrick Smith; and *Inside the Soviet Writers' Union* by John and Carol Garrard. The quotations from Taras Shevchenko are taken from *Taras Shevchenko* by Maxim Rylsky and Alexandr Deich, translated by John Weir.

. . .

"A full interpretation of what happened," wrote Yuri Shcherbak in 1987, "is a matter for the future, perhaps the distant future. No writer or journalist, however well informed he might be, could do that today. The time will come, I firmly believe, when the Chernobyl epic will appear before us in all its tragic fullness."

Who could have predicted in 1987 quite how quickly that distant future would arrive? Or that Shcherbak himself, then a doctor and an amateur writer, would by 1990 be the minister for the environment in an independent Ukraine? In collecting the material for what follows, I have taken advantage of the 1989 declassification of information relating to atomic power, the release of all those imprisoned and access to certain documents and reports that were hitherto secret—the judges' summing up at the trial held in Chernobyl in 1987, the report of the State Committee for Safety in the Atomic Power Industry written in 1990, the report of the Ecological Commission of the Supreme Soviet, the protocols of the Politburo's hitherto secret medical commission set up after the accident, and the numerous reports published by the International Atomic Energy Agency in Vienna.

Shcherbak was undoubtedly right to describe the story as "epic," and one of my most difficult tasks has been to reduce it to manageable proportions. The catastrophe affected several million people. In choosing to tell the story of only a few, I have tried to include some of the essential protagonists and a sample from different groups, but I cannot pretend that I have given every participant his due.

I have emphasized the human side of the story because, in its essence, the accident was caused by human error—but human error of a complex kind. What happened in the fourth reactor early in the morning of 26 April 1986 is now known. What went on in the minds of the operators, and those who took charge after the accident, is not. To understand what happened, however, it is necessary to return to the genesis of the Soviet atomic program in the days of Stalin and Beria; to understand the consequences of the accident we must follow the narrative almost to the present day.

There are certain obstacles that a writer must clear to make the story of the Chernobyl disaster intelligible. The first is atomic physics. With no specialist knowledge myself, I have been alternately daunted by the apparent complexity of the subject and encouraged by its essential simplicity. The narrative is based on the assumption that both writer and reader need only a rudimentary knowledge of how nuclear reactors function. I have included, at the end of the Introduction, an explanation of the terms used for measuring radiation.

The second is a general ignorance about the countries most affected by the accident, Belorussia and the Ukraine. For a large number of people in Western Europe and the United States, it was the accident at Chernobyl that put both countries on the map. This fact in itself affects the story, and if there is any advantage to be seen in the tragedy of Chernobyl, it is in the part it played in re-establishing the autonomy of these countries for the first time in many hundreds of years.

Matched to our ignorance of the two countries is the difficulty we encounter in mastering some of the names. Those like Fomin or Steinberg are easily pronounced and so easily remembered. Others, like Palamarchuk, Davletbayev, Smyshlyaev or Shovkovshytny, are not. The accepted transliteration of the Cyrillic alphabet leaves a plethora of letters in these Slavic names. The wide boulevard in the center of Kiev sounds like "Kreshatic" but is spelled Khreshchatyk.

There are similar problems with first names. The Ukrainian version of Vladimir is Volodomyr, and there are many Ukrainians who take exception to the Russification of their names. I have therefore used both forms of the name. For those Russian names that come close in sound to their English equivalents, I have used English spelling—Alexander rather than Aleksandr; Victor rather than Viktor (in the case of Brukhanov, this was his choice)—but for those where the sound is different, I have used the Russified form—Piotr rather than Peter, Nikolai rather than Nicholas. It is unfortunate that the names Alexander and Lubov, or Luba, are so common among the characters. To assist the reader, I have included a list of leading characters. I have not given patronymics except where they were used in conversation.

The "gigantism" that was such a feature of the former Soviet Union is nowhere more apparent than in the names of institutions that are either referred to by their initials—VNIIAES for the All-Union Research Institute for Nuclear Power Plant Operations, or NIKYET for Academician Dollezhal's Scientific Research Institute for Technical Energy Construction—or by the compacted abbreviations adopted by the

Bolsheviks: Minenergo for the Ministry of Energy and Electrification or Soyuzatomenergo for the All-Union Industrial Department for Nuclear Energy. Where possible, I have avoided these compacted forms and have either used initials or given a simplified form of the full title.

The further I looked into the accident at Chernobyl, the more I came to realize that writing a book on the subject was a task with grave responsibilities. To exaggerate the damaging consequences would confirm fears about nuclear power; to minimize them would not just encourage its development but also implicitly dismiss the pleas for foreign aid made by charitable organizations like the Chernobyl Union, Chernobyl Help or Children of Chernobyl. Ignorant of hematology, immunology and atomic physics, with little knowledge of the peoples and no knowledge of their languages, it seemed at times reckless to pursue my research.

However, as the third part of the story will make clear, it is difficult to see how any Russian, Belorussian or Ukrainian writer could approach the subject with detachment, and even in the West many writers on either the environment or the former Soviet Union are inspired by a partisan zeal. I embarked upon the project with no ax to grind and have done my best to report what happened, leaving it to the protagonists of the different points of view to speak for themselves.

Radiation

Radiation is emitted by unstable atoms from many different sources, including sunlight, granite, and the human body. It can cause harm when its waves and particles penetrate living tissue. There are five types of ionizing radiation:

1. Alpha particles do the most damage but can be stopped by paper, glass or skin.
2. Beta particles cause less damage but penetrate living tissue to a depth of around ten millimeters, leading to beta "burns."
3. Neutrons, uncharged particles of about one quarter of the mass

of alpha particles, are penetrating and damaging, but the circumstances in which people are likely to be exposed to neutron radiation are rare.

4. Gamma radiation comes from electromagnetic rays and can be blocked only by thick slabs of concrete or heavy sheets of lead.
5. X rays, which result from bombarding a metallic surface with energetic electrons, are otherwise similar to and behave like gamma rays.

Atoms regain their stability over a period of time: their radioactivity "decays." This period is different for the various radioactive elements or isotopes and is measured by their "half-life"—the time it takes for half a given sample to decay. In ten "half-lives," virtually all the radionuclides will have decayed. The radionuclide iodine 131 has a half-life of around eight days. Strontium 90 and cesium 137 have half-lives of thirty years. Plutonium 239 has a half-life of 24,360 years.

When ingested or inhaled, different radionuclides are retained in different parts of the body. Iodine accumulates in the thyroid gland, but if the gland is saturated with stable iodine at the time of irradiation or soon after, the radioactive iodine will pass through the body. Strontium replaces calcium in the bones.

The original unit used to measure radiation exposure was a roentgen, after Wilhelm Roentgen, who discovered X rays at the end of the nineteenth century. The amount of radiation to hit the human body is measured in a unit called called the *rad* (radiation absorbed dose), but different types of radiation vary in their biological effects: for example, alpha particles are twenty times as damaging as gamma radiation. The unit devised to reflect this is the *rem* (roentgen equivalent man): for gamma radiation, 1 rad = 1 rem, while for alpha radiation, 1 rad = 20 rem.

New units are now in use. 1 gray (Gy) equals 100 rads and 1 sievert (Sv) equals 100 rems. Since most people involved in the story of Chernobyl used the older units, I have used rads and rems. When sieverts or millisieverts are used in direct quotations, I have inserted the equivalent in rads or rems in brackets.

The same is true when it comes to measuring radioactive contamination of the soil and food. The new unit of measurement is the becquerel (Bq); the old unit is the curie (Ci). I have used the old unit because it is commonly employed in the affected areas. A becquerel is one radioactive disintegration per second: a curie is thirty-seven billion disintegrations per second.

Radiation damages human health in two distinct ways.

1. Non-stochastic, whereby the severity of the effect varies with the size of the dose, and the likelihood of that effect is high once a threshold dose is exceeded. Over 100 rems, a patient is liable to succumb to radiation sickness with nausea, vomiting, diarrhea and changes to the blood. The consequences of larger doses are described in the text. The "median acute lethal dose" of radiation, whereby 50 percent of those exposed would be expected to die within thirty days, is estimated at between 300 and 400 rems. The longer the period over which the dose is received, the less its effect. However, a relatively low whole-body dose measured in rems can accompany a high localized dose to, say, a child's thyroid measured in rads.

2. Below 100 rems, though there may be some blood changes and slight nausea, the dangers are mainly stochastic—that is, it is the *probability* of the effect occurring rather than its *severity* that varies with the size of the dose. A small dose for a larger number of people could lead to the same number of casualties as a large dose for a small number of people. The latent period between radiation exposure and the onset of radiation-induced diseases varies from a few years for leukemia to a few decades for cancer.

PART ONE

THE NEW CIVILIZATION

This fundamental transformation of the
social order—the substitution of planned
production for community consumption,
instead of the capitalist profit-making of
so-called "Western Civilisation"—seems
to me so vital a change for the better,
so conducive to the progress of humanity
to high levels of health and happiness,
virtue and wisdom, as to constitute
a new civilisation.

Sidney and Beatrice Webb
The Truth About Soviet Russia (1942)

1

In the summer of 1942, when the Soviet Union was allied with the United States and Great Britain in a pitiless war against Germany, a young Russian physicist named Georgi Flerov, on leave from the front, traveled to the Tatar city of Kazan. Here were the government ministries and scientific institutes that had been evacuated from Moscow and Leningrad. Anxious to catch up on any developments in nuclear physics, Flerov went to the library of the Soviet Academy of Sciences to read the scientific journals from abroad. He found the journals, which despite the war had reached Russia, but not the articles he had expected. The papers describing the work in progress of Western scientists like Joliot-Curie in France, Rutherford in England or Fermi in the United States were not to be found: there was no mention whatsoever of any research in Flerov's field of nuclear physics.

Flerov realized what this meant. The research had been classified: the Americans were developing a nuclear bomb. He at once wrote a letter to Stalin similar in tone to that which Albert Einstein had written to Roosevelt in 1939. Flerov warned his leader of the theoretical potential for mass destruction inherent in an atomic weapon and urged him to seek the advice of the nation's most distinguished physicists.

Stalin heeded his warning, deciding that the Soviet Union too must develop a nuclear bomb. The scientist he chose to supervise the project was Igor Kurchatov who, in the 1930s, at the Leningrad Physical Technical Institute, had been the first Soviet physicist to achieve nuclear fission. Although not yet forty when Stalin summoned him to Moscow,

Kurchatov had grown a long black beard while recovering from pneumonia in Kazan.

Besides his talents as a physicist, Kurchatov had other qualities that appealed to Stalin. He was a Russian, not a Jew, and, if not a Communist, he could show a pedigree of a proletarian kind: his great-grandfather had been a serf who, after being lost to an industrialist in a game of cards, was sent by his new master to work at an iron foundry on the river Sim in the Urals. To marshal the resources for such a huge undertaking, Stalin appointed as overall director of the atom bomb project the Georgian chief of his secret police, the NKVD, Lavrenty Beria. The terror inspired by his NKVD would ensure the project's secrecy and concentrate the minds of those called upon to provide the resources. Beria also controlled many of the resources himself—the uranium mines worked by prisoners, and the huge network of gulags where many of the best physicists were to be found. In the terrible purges of the 1930s, when anyone with a bourgeois background was a potential enemy of the people and professionals in the tens of thousands were incarcerated on any pretext whatsoever—a chance remark, an unfortunate friendship or the anonymous denunciation of an envious colleague—many eminent physicists had been arrested and transported to labor camps in Siberia. For example, at the Institute of Physics in Kharkov, where vital research in nuclear physics was being done, both the head of the atom-splitting department, Leipinski, and a professor of theoretical physics, Lev Landau (whom Flerov had mentioned in his letter to Stalin), had been arrested.

Beria's solution to the problem was not to order the release of the imprisoned physicists but to build camps within camps where the secret work could be done. Thirty thousand meters of tarpaulin were allocated to make tents for the physicists, and fifty tons of barbed wire to fence them in. A special ministry was formed with an anonymous name to conceal its secret purpose—the Ministry of Medium Machine Building. It was headed by an energetic engineer, Efim Slavsky, who had joined the Bolsheviks under Lenin and had fought in the cavalry under Semyon Budenny in the civil war. Under his direction, everything asked for by Kurchatov was wrung out of an economy shattered by war. An experimental reactor was built in a secret laboratory in the woods outside Moscow, where there were flats for the physicists and a villa for their director. Forty kilometers from Moscow a whole town was devoted to the nuclear project, including a camp for German physicists captured in the closing months of the war. It was a town with no name, unmarked on maps, and bare of roadsigns; by the same token Kurchatov's name was never mentioned; his code name was Borodin, while to his friends he was simply "The Beard."

On 14 July 1945, Stalin met Churchill and Truman at Potsdam, a suburb to the west of Berlin. Two days later, the Americans successfully detonated the world's first atomic bomb in the New Mexican desert at Alamogordo. On the 24th Truman told Stalin that the United States had a new weapon of extraordinary destructiveness. Stalin showed no special interest, but that night he cabled Beria to hasten the development of a Soviet bomb. When he returned to Moscow he demanded weekly progress reports from Beria. In August he learned that the Americans had not only built an atomic bomb but were prepared to use it—obliterating the Japanese cities of Hiroshima and Nagasaki and bringing the war in the Pacific to an end.

An acute sense of urgency spurred on Kurchatov and his team. Whatever personal misgivings any may have had about Stalin or the Communist system, there was a real fear that the struggle against the Germans might be followed by a war with the Americans. The development of the bomb became the top priority of the Soviet state. No expense was to be spared, and the authority of Kurchatov, as scientific director, became greater than that of a minister. In the autumn of 1946, he was authorized to hire up to thirty-seven thousand workers to speed up the construction of the facilities he required. Where laborers were lacking, convicts took their place. Five thousand prisoners were transported from prison camps in Siberia to work on the facilities at Mayak, near Chelyabinsk in the Urals. Where zeal was lacking, Beria used terror.

It was a style that others would emulate. Even Kurchatov, cultivated and humane though he was, made maximum use of his good standing with Stalin when dealing with recalcitrant government ministers. With his scientific colleagues, Kurchatov was more easygoing. They could argue with him—even lose their tempers, leave the room, slam the door—with no adverse repercussions. Although rumored to be anti-Semitic, he had several Jewish scientists on his team. He employed one,

Boris Dubowski, to take charge of safety at the town with no name south of Moscow, even though Dubowski had no formal qualifications, and he personally sought Stalin's authorization to allow Dubowski to travel to Chelyabinsk. Another Jewish physicist who had a sharp sense of humor was allowed to mock and mimic the most eminent ministers and academicians.

Although Kurchatov directed the project, he was assisted by a team of top physicists like Khariton, Kikoyin, Kapitsa, Tam and Tam's ablest pupil, Sakharov. He also benefited from the intelligence provided by the Soviet agent Klaus Fuchs in the United States. On 25 December 1946, in his secret laboratory on the outskirts of Moscow, Kurchatov and his team achieved their first chain reaction. However, they were still some way from a bomb. For this they required plutonium, a product of the fission of uranium. This plutonium could only be produced in sufficient quantities in the atomic reactor being built at Mayak.

2

For the design of this reactor, Kurchatov had turned to one of the country's leading engineers, Nikolai Dollezhal. A small, unassuming man, Dollezhal was typical of the "bourgeois specialists" whom the Bolsheviks had employed to develop the industrial capacity of their Communist state. He came from a professional family in the Ukraine. His grandfather had been a Czech engineer employed by an Austrian company to build a bridge over the Dnieper for the railway from Kursk to Odessa. Nikolai was baptized into the Orthodox church, sang in the church choir as a boy and spent his holidays on his Russian grandmother's crumbling, neglected estate. His interests, like those of his parents, were in music and literature, not politics: the dramatic events of 1917 passed him by.

Graduating from the prestigious Technical University in Moscow—formerly the Imperial College—in 1923, he worked both in the power and chemical industries as both a designer and an engineer. He never became a Communist: although frequently invited to join the party, he always made the excuse that he could not claim to be a Marxist since he never had time to read Marx. Arrested in 1930 on suspicion of complicity in the "Industrial Party" conspiracy, which was to lead to the first of Stalin's show trials, Dollezhal was released for lack of evidence and went back to work in the power industry in the Ukraine.

By 1945 Dollezhal had moved to Moscow. He was a member of the government's technical advisory council and director of an institute that he had founded to design machinery for the chemical industry. It was in this capacity that he was approached by Kurchatov and invited to join his team.

The two had met before at a game of tennis in Leningrad in the 1930s. Now the task at hand was more serious: to design a reactor that would produce plutonium for the bomb. This was to be not just the primary but the only purpose of the design Kurchatov was asking for, which also had to take into account what could be produced by Soviet industry at the time. Both zeal and terror could extract extraordinary achievements from human beings, but time was short; they could not wait to develop technologies to match the optimum design.

Within these constraints, and with astonishing speed, Dollezhal planned a reactor to meet the requirements of Kurchatov and his team. With the resources provided by Slavsky's Ministry of Medium Machine Building, the reactor was built in the utmost secrecy at Mayak. On 10 June 1948 it was commissioned. A little over a year later, in July 1949, the first Soviet nuclear explosion took place at Ustyurt in the desert of Kazakhstan. Three months later, on 23 September 1949, they successfully detonated the first Soviet atom bomb. "Now," said Kurchatov, "we have our atomic sword and can start thinking about peaceful uses for the atom."

3

From the first years of the twentieth century, when it had been discovered that by splitting the atom mass could be converted into energy, it had also been understood that heat generated by this nuclear fission could be converted into electrical power. A decade before, this had been only a theoretical possibility; now, with a functioning reactor at Mayak, it became a practical one, which Kurchatov tackled with his usual zeal.

No one at this time stood higher in Stalin's estimation. Kurchatov was not only given every decoration worthy of his achievements, including the first Order of Lenin ever awarded, but also had bestowed upon him the material rewards reserved for the favored few. Besides the spacious house in the woods by his secret laboratory, called the "Forester's Cabin," he was now given a villa in his beloved Crimea. Dollezhal, too, was rewarded with a dacha—a pleasant house in the woods of Zhukovski outside Moscow, where he found among his neighbors the composer Shostakovich, the cellist Rostropovich; and the physicists Flerov, Sakharov and Tam. Kurchatov was a frequent visitor, often playing with Dollezhal's daughter in a little hut in the garden: he had no children of his own. There they would also discuss Kurchatov's plans to turn the nuclear sword into a nuclear plowshare, not just to generate electricity but to drive submarines, icebreakers, locomotives and even airplanes.

In 1949 Kurchatov sought Stalin's approval to build an experimental nuclear power station at the town with no name south of Moscow where the captured German physicists were held. Dollezhal was part of the scientific team; he designed the turbines while Kurchatov himself chose the reactor. It was a graphite-moderated, water-cooled model similar to that which was already up and running in Kurchatov's own laboratory and at Mayak. The project was approved, but Stalin did not live to see it function. He died in 1953, and in December of the same year Beria and his closest associates were shot.

Kurchatov did not fall with his erstwhile patrons; quite to the contrary, he was both liked and admired by Nikita Khrushchev, who subsequently came to power. The experimental nuclear power station at the town with no name—the first in the world—went on line on 27 June 1954. There was no fanfare for this achievement of Soviet science and technology because the whole Soviet state remained in the grip of the obsessive secrecy of Stalin's time. It was only two years later, after his denunciation of Stalin at the Twentieth Party Congress, that Kurchatov was named as the founding father of Soviet nuclear power. Borodin could now drop his mask; "The Beard" became familiar outside his small circle of colleagues and friends. The laboratory outside Moscow was named after Kurchatov, and the secret city was finally given a name: Obninsk. When Khrushchev, with Bulganin, paid a state visit to Britain in 1956, Igor Kurchatov went with them and spoke to British scientists about Soviet nuclear power and the bright future for the peaceful atom.

In the Soviet Union itself, not everyone shared Kurchatov's vision. Khrushchev's chief scientific adviser, Vladimir Kirillin, argued vigorously against atomic power because of the unsatisfactory return on the investment required. In contrast to the hydroelectric power stations planned at Bratsk on the Angara River, which would produce forty-five hundred megawatts of electricity, the experimental power station at Obninsk produced only five megawatts—insufficient for the plant's own requirements. The nuclear power station planned at Beloyarsk was to produce a mere fifty megawatts.

In his struggle with Kirillin, however, Kurchatov had the backing of the Ministry of Medium Machine Building. Beria might be dead, but the empire he had founded remained a state within a state, incorporating the burgeoning defense industries and with loyal allies in the Ministry of Planning, Gosplan, and the KGB. It could always play the trump card of the ideological imperative: when it became known that the Americans were developing a new type of reactor cooled by water held under pressure so that it could not boil, it was not difficult to persuade the Central Committee that the Soviets must do the same. Authorization was given for the construction of a new power station using a reactor of this kind at Novovoronezh, about thirty miles south of the city of Voronezh.

4

The man chosen to head the development of the pressurized water reactor was a scientist who had worked with Kurchatov throughout his career, Anatoli Alexandrov. Like Dollezhal, he too was a "bourgeois specialist": his father had been a judge in the Ukraine until the Revolution of 1917, when he had changed his profession to that of a schoolteacher: a new species of justice required judges of a different kind. Showing an aptitude for physics, the young Anatoli had taken a post at the Roentgen Institute in Kiev. In due course, his work there attracted the attention of Abram Ioffe, who offered Alexandrov a job in his institute in Leningrad, working alongside the young Igor Kurchatov.

· · ·

In developing a pressurized-water reactor, Alexandrov faced a formidable task because the requirements for the large pressure vessels tested Soviet technology to its limits. As a result, the construction of Novovoronezh was delayed over and over again by problems with the reactor, and the Soviet government eventually decided to revert to conventional, fossil-fueled power generation.

When Kurchatov got wind of this, he drove straight from his institute to the Kremlin and there insisted that the Central Committee and Council of Ministers reconsider their decision. He argued forcefully that the Soviet Union must be in the vanguard of nuclear science. Was it not the first society in human history to be based upon scientific principles? Could they allow it to be shown that the world's first Socialist society, the new civilization, built according to the precepts of Marx and Lenin, was incapable of matching the technological advances made in the decaying, capitalist West? Had not Lenin defined communism as Soviet power plus electrification? Then how could the application of science par excellence to the generation of electricity be abandoned for the coal and water that had been in use since feudal times?

Kurchatov had his way: the resolutions were rescinded. When asked by a skeptical minister when they could hope to see tangible benefits from nuclear power, Kurchatov replied that for twenty-five or thirty years they should regard the program as no more than an expensive experiment. Only then would be benefits become apparent; they would have to wait until around 1985.

The debate about nuclear power focused on cost, not safety. Safety was never an issue; yet during these deliberations some members of the Central Committee, the military high command and the Ministry of Medium Machine Building had known that in 1957 there had been a catastrophic accident at Mayak. Radioactive waste had been stored a mile or so from the plants, in concrete tanks lined with stainless steel. These tanks had been cooled by water piped through their walls. In 1956 leaks of this coolant had been noticed, but nothing had been done to stop them. It had been assumed that the waste was stable, but it had dried out and, when ignited by a spark from a control device, chemicals had exploded—blowing the lid off the tank and sending twenty million curies of radioactive elements into the atmosphere.

Most of the heavy particles had fallen back to the ground in the vicinity of the tanks, but around two million curies of lighter particles had been carried off by a southwesterly wind toward Sverdlovsk. No deaths were ascribed to the accident, but more than 10,000 people were evacuated and 250,000 acres of agricultural land laid waste.

Two further accidents were to follow at Mayak: the first, when radioactive waste was dumped directly into the Techa River; the second, when the highly radioactive waters of an artificial lake used to cool the reactors was whipped up by a cyclone and spread over a large area of land around Kyshtym. Eight thousand people living on the banks of the Techa River were evacuated and many thousands of acres were made barren.

All these accidents were hidden from the outside world. No one broke the code of *omertà*. It was not just the fear of the KGB but the esprit de corps of those who worked for the military-industrial complex under the aegis of the Ministry of Medium Machine Building. Patriotism, too, played its part. Few doubted their government's propaganda that the Americans were preparing for a war to obliterate the new Socialist civilization. Irene, the daughter of Pierre and Marie Curie, and her husband, Jean-Frédéric Joliot-Curie, a pioneer of nuclear fission, were both Communists and on international questions followed the party line. So too did a large number of Western intellectuals like Louis Aragon and Jean-Paul Sartre; there was therefore no reason for the Soviet scientists to doubt the existence of saboteurs and spies. Closed cities like Mayak and Obninsk were surrounded by barbed-wire fences and guarded by whole regiments of special troops.

This esprit de corps was sustained by the favorable conditions within the perimeter. There were theaters and cinemas far superior to those found in other provincial cities. The flats were better built and the shops stocked with goods that were unobtainable elsewhere. Absolute discretion was maintained with both stick and carrot: no one wanted to betray his country, go to prison or lose his job.

It was also noted that the accidents at Mayak had nothing to do with the reactors themselves. With such a new and advanced technology, it was impossible to anticipate every eventuality: the most important thing for the scientists was to learn from these mishaps about the effects of radiation, not so much to prepare them for further accidents but for the aftermath of the anticipated atomic war. An institute of biophysics was set up next to the Kurchatov Institute in Moscow, which had laboratories at Mayak. A young radiobiologist, Leonid Ilyn, wrote his thesis on what he had learned after research in the area around Kyshtym on the absorption of radioactive strontium in meat.

Kurchatov's personal physician, a young woman called Angelina Guskova, at Mayak from 1948 to 1958, learned from the accident how best to treat victims of radiation sickness. In 1957, at the age of only fifty-four, her principal patient, Igor Kurchatov, suffered a stroke. There was no specific link between this and the dose of radiation that he had inevitably accumulated in the course of his life, but the struggle to arm his country with an atomic bomb had meant years of incessant labor with great stress and little sleep. Even after the stroke he continued to work from a sanatorium, where three years later he suffered a second and fatal seizure while sitting on a bench next to a fellow physicist, Academician Khariton.

Kurchatov died in the knowledge that his work was largely done. Not only was the Soviet Union armed with nuclear weapons, but it was committed to the development of nuclear power. From the small group of young scientists whom he had gathered around him in 1943, there had grown an enormous empire. His institute in Moscow employed ten thousand people; Obninsk was a city of one hundred thousand. Now not just the old institutions, like Ioffe's Physical Technical Institute in Leningrad or the Roentgen Institute in Kiev, but every major university and technical college had its department of nuclear physics. The armed forces had their own facilities, from the testing grounds in the deserts of Kazakhstan to the dockyards at Komsomolsk, where reactors were fitted into nuclear submarines. His service to his nation had been outstanding and was recognized by those in power: Nikita Khrushchev was one of the pallbearers who carried his coffin to his grave.

5

The man chosen to succeed Kurchatov as sovereign of his atomic empire was his companion from the early years in Leningrad, Anatoli Alexandrov. Now aged fifty-seven, he was a man whose scientific skills were augmented by a talent for administration and a natural authority—no doubt inherited from his father, the czarist judge. He was tall with a huge hairless head, pointed ears, an aquiline nose and a commanding manner. Already an academician, decorated with medals for the defense of Leningrad, Stalingrad and Sevastopol, the winner of an order of the Red Banner, an order of the October Revolution, three Heroes of the Soviet Union and four state awards, as director of the Kurchatov Institute Alexandrov now gained patronage of an unparalleled kind. It was he who now decided subordinate appointments and allocation of funds, research projects and trips abroad. As editor in chief of *Atomenergo* he controlled the publication of all the findings in the field of nuclear physics, and as director of the Kurchatov he could take advantage of the well-established practice of publishing papers written by his subordinates under his name.

Alexandrov was also now a leading member of the unacknowledged aristocracy of Soviet society, into which only the inner circle of political leaders were admitted, along with a few favored scientists, writers, artists and musicians. As an academician, he became entitled to a dacha and a larger flat in a better building; a further emolument was added to his salary as the director of an institute. With his family, he had access to special recreational facilities and medical care; his children could attend elite schools and coveted institutes of higher learning. There was a black chauffeur-driven Volga to take him from his flat to the Kurchatov Institute or to the beautiful Neskuchny Palace, which housed the presidium of the Academy of Sciences.

Alexandrov was not alone among Kurchatov's colleagues to be rewarded in this way. Dollezhal, too, was made a member of the Academy of Sciences and was now the director of his own bureau for designing atomic reactors, the Scientific Research Institute of Technical Energy Construction, or NIKYET. Never fond of each other, as each had been of Kurchatov, the two leaders of the nuclear industry were now further separated by the distance between their institutes and the vast armies of underlings each had at his command.

The mutual antipathy of these two leaders of the atomic community did little to help solve the practical problems that, despite the growth of the theoretical institutes, continued to bedevil the industry. At the start-up of a new reactor at Mayak, Alexandrov himself had to take over control from an operator to prevent an accident. At the Kolski nuclear power station an operator happened to notice steam coming out of a pipe. The reactor was shut down, the pipe examined and a crack found in the molded seam. Further checks were made, and twelve additional seams, which the inspectors had certified as sound, were discovered to be faulty.

Deficiencies of this kind were particularly dangerous in the pressurized-water (VVER) reactors and the projected fast-breeder reactors, where the coolant was liquid sodium, which, if brought into contact with water as a result of a ruptured pipe, would lead to both an explosion and a fire. The later VVER reactors had to be built with complex and expensive containment structures to prevent the spread of hazardous radiation in the case of an accident.

Given the difficulties they faced in developing the VVER and fast-breeder reactors, it was a comfort to Alexandrov and the Ministry of Medium Machine Building to have the tried and tested design of the very first reactors at Mayak and Obninsk. If the VVERs were thought of as the gazelles of the industry, this old design was seen as the workhorse. It had proved so safe and reliable that there seemed no need for an expensive containment structure; and the industrial base required to build it had been in existence since the war. All that was required to increase its capacity was to increase its size.

Following this reasoning, there emerged from the drawing boards of Dollezhal's institute plans for a "high-powered, boiling, channel-type reactor"—the RBMK—which would generate one thousand megawatts of electrical power. Although it used the same graphite moderator, uranium fuel and water coolant as the early prototypes, there had been developments in the design. For example, the turbines at Obninsk had been driven by steam in a separate circuit, which had led to a considerable loss of thermal energy at the point where the heat was exchanged. At Beloyarsk, therefore, and in subsequent reactors of this kind, the steam came straight from the fuel channels of the reactor itself.

The advantage of this design was not just the increase in efficiency but the more modest technological demands. The temperatures of the water and the pressure of the steam in the pipes had been substantially reduced, and the only defenses considered necessary were watertight walls around the reactor and the circulation pumps, so that any leaks of radioactive steam or water could be fed into the tanks, known as the "bubbler pools," beneath the reactor.

Three airtight cylindrical containers filled with gas, water and sand surrounded the reactor itself. Between the reactor and the bubbler pool below there was a thick concrete floor, and above the reactor there was a neutron shield made of steel and concrete. It was seventeen meters in diameter, three meters thick, and perforated with holes to enable the fuel and control rods to enter the reactor. One of the advantages of the RBMK design was that it could be refueled without being closed down.

These safeguards, however, were to protect the operating personnel from the harmful radiation emanating from the reactor, not to contain a potential explosion. Although the graphite blocks surrounding the fuel channels were combustible, the worst hazard imaginable was the rupture of one or possibly two of the fuel rods, which could at most cause a localized leak of radioactivity. The reactor itself could always be kept stable by the 211 boron control rods, which when lowered into the reactor absorbed the neutrons and either slowed the rate of fission or brought it to an end.

In the early 1960s, approval was given for the construction of two RBMK-1000 reactors outside Leningrad. So confident were Alexandrov and Dollezhal of this design that they proudly described it in 1971 to the Fourth International Conference on the Peaceful Uses of Atomic Energy in Geneva. Even though the Leningrad reactors were not yet operational, the go-ahead had been given to build further reactors of the same type in other parts of the Soviet Union—among them Ignalina in Lithuania and Chernobyl in the Ukraine.

The few misgivings expressed about the design, both inside and outside the Soviet Union, were either dismissed, overlooked or ignored. The Americans had used graphite-moderated, water-cooled reactors to produce plutonium, but not for civil power. British graphite-moderated reactors were cooled by CO_2. The lack of a containment structure, and the danger of a "positive void coefficient" were among the seven reasons given by the British Atomic Energy Authority as to why a RBMK-type reactor could not be licensed in the United Kingdom. There was also a considerable margin between theory and practice; for example, it took eighteen seconds to lower the control rods into the reactor core, although the physicists had said that it should take three.

The enormous size of the active zone of the RBMK-1000 reactors worried Boris Dubowski, the man whom Kurchatov had appointed to head the department of nuclear safety at Obninsk. At a meeting in 1976 called to discuss safety at the RBMK reactor built near Kursk, he suggested that extra boron control rods be installed into the lower part of the reactor. In this he was supported by Alexandrov, and it was decided to recommend to Dollezhal's bureau that this be incorporated into the design. But even though the measure had Alexandrov's blessing and the chief nuclear safety inspector was present at the meeting, the modifications were never made. Like so many other measures during this period, which was to be called the era of stagnation, the idea traveled sluggishly through the clogged arteries of the obese Soviet administration, moving from department to department and committee to committee in the vast bureaucracies of NIKYET and the Ministry of Medium Machine Building, where a combination of the expense involved in making the changes and the apparent safety record of the RBMKs ensured that nothing was done.

In addition, those who had called for the changes had other things on their minds. Dubowski himself, following an accident at Obninsk for which he was technically responsible, lost his position as head of the nuclear safety department, while the year before Alexandrov had been appointed president of the Academy of Sciences. This was a majestic achievement that not only recognized his accomplishments as a scientist but also revealed the continuing prestige of the military-industrial complex in the collective mind of the Central Committee. Unlike an atomic scientist in the West, Alexandrov was no mere back-room boffin who came up with clever ideas; he was a scientific leader, a general in command of battalions dedicated to the Bolshevik cause. New discoveries were made both to serve and to defend the new Soviet civilization, and with the glory came the usual perquisites of power. No longer was Alexandrov driven to and fro in a mere Volga; like General Secretary Leonid Brezhnev himself, he became eligible for a Zil. But unlike the party leader, who looked like a gorilla, Alexandrov, the judge's son, had the haughty demeanor of a true aristocrat; and when he descended from his huge limousine to walk up the elegant staircase of the Neskuchny Palace, where he now reigned supreme, it was as if to the manner born.

6

Preoccupied now with his wider duties as president of the Academy of Sciences, Alexandrov delegated many of his powers as director of the Kurchatov Institute to his first deputy director, Valeri Legasov. A much younger man than Alexandrov, Legasov shared his leader's zeal for the cause of their Communist country; but he was no "bourgeois specialist," having been born after the Revolution into the privileged elite of Soviet society. His father, Alexsei Legasov, the son of an Orthodox priest, had been head of the ideological department in the secretariat of the Central Committee, which, like the Holy Inquisition in the Roman Catholic church, decided what conformed to the truth faith. As a student, Valeri served on the committee of the Communist Youth organization, Komsomol, and in 1951—impatient to do something dramatic for his country—he led brigades of students from the Mendeleev Institute, where he studied, to bring in the harvest in Siberia.

The work Valeri had first embarked upon at the Kurchatov Institute followed from his graduate research into the chemistry of noble gases. Drawing around him a team of able scientists like Vladimir Klimov and Nikolai Protzenko, he and his team rapidly made discoveries of a theoretical nature that had practical applications in industry and defense. His extraordinary capacity for hard work, for taking risks, for understanding and solving the problems that arose in the course of their research made Legasov the natural leader of his group. The position held by his father in the Central Committee gave him a confidence no ordinary Soviet scientist could afford: he would take risks in his research, and also mock the pomposity of the most eminent party leaders who visited the Kurchatov. He was a wonderful raconteur, and because of his access to foreign books and journals, which he would sometimes lend to his colleagues at the Kurchatov, he had a breadth of culture far greater than that of the general product of the Soviet educational system.

Legasov rose rapidly in the hierarchy of the Kurchatov Institute: from senior researcher to head of the lab, from head of the lab to head of the department. Loyal to the system in which he fervently believed, and with ambitions beyond the realm of science, he served for a period as the institute's Communist party secretary. This political zeal, combined with his scientific achievements, led to further promotions. For his work in applying fluoride chemistry to military technology, he was made a corresponding member of the Academy of Sciences. When Alexandrov was appointed its president in 1975, Legasov was his choice as first deputy director. From then on he became responsible for running the Kurchatov Institute, a position of power that made him both enemies and friends. To old friends like Protzenko, he became remote, distanced not just by his eminence, but by a coterie of ambitious sycophants who fawned on him to promote their own advancement.

All the patronage that had been Alexandrov's was now in Legasov's hands. In the allocation of funds he made enemies, such as the physicist Yevgeny Velikhov, who led the team doing research into nuclear fusion. To Legasov, this was a dream that consumed billions of rubles from their budget; to Velikhov, Legasov was a chemist who, though his work may have led him into physics, was not qualified to rule on questions of this kind.

Legasov also had to decide the petty but all-important question of which scientists should be sent to congresses and symposia in the West. Here race played a role; the Jewish scientists were considered too assiduous in putting themselves forward for perks of this kind. Legasov, who was Russian, held them back: by reason of equity, according to the Russians; from prejudice, according to the Jews. There was also rancor among many of his colleagues that Legasov had risen so far and so fast. The influence of Brezhnev was thought to be behind his promotion; Legasov remained the golden boy of the Central Committee. It was unquestionably satisfying for the party leadership to see the success of one of their own—to approve his appointment as a professor at the Physical Technical Institute of Moscow University, and in 1980, when he was only forty-four years old, to make him a full member of the Soviet Academy of Sciences.

7

It was not just the patronage of politicians, but his adoption by Alexandrov that made Legasov his heir apparent. Already over seventy when he was made president of the Academy of Sciences, it was beyond Alexandrov's capacity to perform the functions that this office entailed, as well as direct an institute of ten thousand physicists and supervise atomic power throughout the Soviet Union.

In 1979 a crisis arose in this domain because of an accident outside the Soviet Union. On 29 March at 4 A.M. at a nuclear power station at Three Mile Island in Pennsylvania, a fault in the boiler feed system led to the lifting of a safety valve in a reactor primary-cooling circuit. This led to a loss of core-cooling water. Unaware of what had happened, the operators in the control room shut off the emergency cooling system, which had automatically come into operation. Water in the core of the reactor began to boil; the uranium fuel rods overheated and finally ruptured. When this critical situation was discovered two hours later, water was pumped into the reactor to restore core cooling. By the time the operators regained control of the reactor, the core was badly damaged, in part melted and slumped down, and the cooling water was highly radioactive. Little radioactivity escaped into the atmosphere, but thousands fled the vicinity of the reactor, and many people throughout the world lost faith in the assurances they had been given that nuclear power was safe.

Although it was tempting for the Soviet leaders to gloat over the discomfiture of their rivals in America, they were aware of the skeletons hidden in their own closets. Nor did they wish to alarm their own people about the potential dangers of nuclear power. Thus instructions were given that little about the accident was to appear in the Soviet press.

However, the facts were known to those academicians, like Nikolai Dollezhal, who had access to foreign journals, and later in the year, assisted by an economic specialist, Yuri Koryakin, he wrote an article on Soviet nuclear power that was published in the magazine *Kommunist*. He began with a description of the development of the industry to date and continued with assurances about its safety. "Designs of nuclear power stations," Dollezhal wrote, "take into account any kind of emergency situations, even hypothetical ones, so that later they cannot endanger the personnel at the plant or the environment." Soviet scientists, he insisted, "because they do not have any other interest but the interests of their people," always make technical decisions based "primarily on human considerations"—unlike, he implied, the capitalist Americans, who were only interested in profit.

Nevertheless, there were problems, notably with the processing of spent fuel and the disposal of nuclear waste. In words laden with significance for those who knew about the accidents at Mayak, Dollezhal reminded his readers that the handling of waste remained "a major problem. . . . This is why places for the regeneration of nuclear fuel are located far away from industrial areas and populated settlements." On the other hand, most of the planned nuclear power stations were to be built in the European part of the Soviet Union—west of a line drawn from the Volga to the Baltic, an area that contained 60 percent of the country's population. This meant that they were encroaching upon the most valuable land and damaging the environment. "The current principle of deployment," he warned, "could very quickly, in our opinion, lead to the exhaustion of the ecological capacities of the region."

Dollezhal and Koryakin estimated that the land required to build fifty nuclear power stations could grow enough food for several million people and that the amount of water lost through evaporation in the already fertile regions was in danger of damaging the ecological balance. The solution to the problem, wrote Dollezhal, was to build huge clusters of new nuclear power stations in the remoter areas of the Soviet Union—Siberia, the Arctic and the Far East—where water was plentiful and the land barren.

Dollezhal's article was immediately refuted by Alexandrov. Taking the unusual course of calling a press conference to which Western diplomats and foreign journalists were invited, Alexandrov insisted that atomic energy was completely safe. Dollezhal, he said, merely designed reactors, and his colleague Koryakin was an economist; how could either of them feel qualified to pronounce on nuclear technology? The suggestion that had been made by Dollezhal, and also by the nuclear physicist Academician Per Kapitsa, was dismissed by Alexandrov as absurd.

Alexandrov's reaction was not just the expression of an old man's pride. The dispute between the academicians reflected differences between party leaders about how the development of the Soviet energy industry should proceed. Energy had always been seen as a measure of the strength of the Communist system itself, and the country's appetite for power had grown at an accelerating rate since the end of the Second World War. Generating capacity, only 1.2 gigawatts in 1920, had risen to 11.2 gigawatts by 1940, reaching 295 gigawatts by 1983.

Yet despite these dramatic increases, demand for power continued to exceed supply. Not only was Soviet industry profligate in its use of power, but the chief reserves of fossil fuels were to be found in Siberia, thousands of miles from the centers of industry in the European part of the Soviet Union. Added to this was the need for the Soviet Union to provide energy for its satellite countries in Eastern Europe, none of which had satisfactory sources of their own.

The large coal mines in the Donbas region of the Ukraine and in the fields south of Moscow were rapidly becoming exhausted; the coal that remained was in deep seams that were expensive to exploit. Bringing coal from Siberia to replace these stocks already tied up 40 percent of the freight on the country's railways; to build more coal-fired power stations could only make things worse.

There were alternative sources of energy in oil and gas, both of which could be piped from Siberia, where reserves were plentiful, but these were also the only commodities for which the Soviets could find a ready market abroad. With recurrent shortages of grain from poor harvests that had to be made up by purchases in hard currency on the international market, Soviet planners became loath to squander this precious asset on their own energy requirements. It seemed much more sensible to provide for the shortfall in energy by a rapid expansion of nuclear power.

It was for this reason that the full authority of the Central Committee and the Soviet government came down on the side of Alexandrov. Dollezhal's article, which some had seen as the beginning of a debate, turned out to be a flash in the pan. No further criticisms were published; Alexandrov, assisted by Legasov, made sure that the journal *Atomenergo* rejected any articles that dealt with the question of safety. In the numerous papers included in the twelve issues published between 1975 and 1987, none even referred to real or potential accidents in the industry.

Privately, Legasov was concerned about the safety of various installations—chemical plants and thermal as well as nuclear power stations—in the event of war. The Israelis' bombing of a Soviet-built reactor in Iraq had shown that there were potential nuclear hazards in a conventional war. Power might be cut off from the servomotors for the control rods and the pumps of the cooling system before the reactors could be shut down. However, he had no reason to suppose that under normal operating conditions the RBMKs were not entirely safe, and in November 1985 he co-wrote an article in the magazine *Priroda* reassuring its readers of the safety of the RBMK reactors.

This was now the party line. The eleventh Five-Year Plan, launched by President Brezhnev in December 1980, accepted Alexandrov's view that "the entire deficit in the fuel and power balance should be covered by a substantial expansion of atomic power." In 1981, at the Twenty-sixth Party Congress, it was stated that any expansion of electrical production in the European part of the USSR would be in the nuclear and hydroelectric sectors, and Yuri Andropov, Brezhnev's successor as general secretary, confirmed to the presidium of the Central Committee that "the future of our power industry lies first and foremost in the use of the latest nuclear reactors."

8

Alexandrov had triumphed and, given his present preeminence and past achievements, it was only appropriate that his eightieth birthday on 13 February 1983 should be a cause for several celebrations. In the Kremlin, a ninth Order of Lenin was awarded by Yuri Andropov himself. At the Academy of Sciences their president was acclaimed by a gathering of the nation's most distinguished citizens; scientists, statesmen, even astronauts were there. His family gave a private party at home, while Legasov and his colleagues at the Kurchatov prepared elaborate festivities in the institute's House of Culture—first a concert and speeches in the auditorium to which four hundred and fifty were invited; then a more select party in the restaurant.

Someone noticed that there were eighty steps leading up from the entrance of the House of Culture to the auditorium, and so on each step they recorded the achievements appropriate to that year, and over the elevator—for some of Alexandrov's old friends like Academician Khariton could hardly be expected to climb the eighty steps—they had painted a huge paw like that of a gorilla with the words, "A hairy hand did not push him to the top"—a reference to the way in which the late General Secretary Leonid Brezhnev had given preference to his family and friends.

Accompanied by his wife and family, and with his loyal lieutenant, Valeri Legasov, at his side, Alexandrov took his place before the podium, and the festivities began. There were speeches full of praise for the distinguished physicist's achievements; and even ribald jokes, as when Alexandrov's wife complained that even now, at their advanced age, he pestered her with his attentions. This brought cheers from the audience of friends and admirers. It was his taste for red meat and vodka that did it! Then to round off the evening there was a competition: a box of chocolates for the first person to interpret the signaling flags that old friends from Alexandrov's days in the navy had strung out as a backdrop on the stage.

It was shameful: the old men could not remember. But then a little boy, Alexandrov's grandson, sitting with his family on the front row, stood up and read what it said: "Keep going along the same lines for another eighty years, Anatoli Petrovich!" He was right. There were more cheers. The boy went forward to collect the box of chocolates and was promised as an extra prize one of the three remaining hairs from his grandfather's head.

PART TWO

CHERNOBYL

For man must strive, and striving
he must err.

> Johann Wolfgang Goethe
> *Faust, Part I* (1801)

II

1

In the late 1960s, when it had become apparent that the demand for energy in the European part of the Soviet Union could not be met without an increase in nuclear power, the decision had been taken by the Council of Ministers of the USSR to end the monopoly of the Ministry of Medium Machine Building and permit nuclear power stations to be commissioned and operated by the Ministry of Energy and Electrification. There was no change to the status of the RBMK-1000s that had already been built in Leningrad and Ignalina; and anything to do with nuclear power remained classified. The power stations that used nuclear reactors would have their own detachments of guards and departments of classified information run by the KGB, but they would no longer be confined to the military-industrial empire ruled by the aging Efim Slavsky.

Thus when a power station was to be built in the Ukraine, the choice of a site and of a director was left to the Ukrainian minister of energy and electrification. This man, Aleksei Makukhin, advised by experts from the Kurchatov Institute and NIKYET, Dollezhal's bureau, looked for a site close to Kiev, the capital of the Ukraine, where the electrical power was required.

To the south of the city lay the valuable plains of black earth that, before the collectivization of agriculture by the Communists, had provided enough grain to feed the whole Russian empire and still export a surplus through Odessa. Fifty miles north, however, at the top of the huge artificial lake made by damming the Dnieper in the 1930s—one of the early triumphs of socialism—there was a region known as the Polessia. Here the Dnieper was joined by the Pripyat River where it emerged from the Pripyat marshes—a huge area of swamps, lakes and forests stretching five hundred miles into Belorussia and to the west to the Polish border. It was a neglected area with sodden, sandy soil, populated largely by peasants who in the 1930s had been forced into state or collective farms. They spoke their own dialect and, besides working for the state, cultivated their own plots of land.

The administrative capital of the area was a small town called Chernobyl, built on a spit of elevated land beside the Pripyat. Founded in the twelfth century by Prince Strezhiv from Kiev, it had developed, by the early 1960s, into a small regional center with a hospital, a polytechnic institute, an agricultural college and a music school, as well as the cultural facilities found in any Soviet county town—a cinema, a library and a House of Culture. There were a few small industrial enterprises, some food-processing plants, and a shipyard to repair the river boats from the Pripyat and the Dnieper.

The site chosen for the new power station was twenty kilometers to the north of Chernobyl, close to the Belorussian border, where the Pripyat River was crossed by the railway line that ran from Ovruc to Chernigov. It was to be named after Lenin, and the land was surveyed not just for the huge RBMK reactors and turbine halls, but for a new town to house the workers. In all, twenty-two square kilometers of land were transferred by government decree from the sovkhoz and kolkhoz—the state and collective farms—to the Soviet Ministry of Energy and Electrification.

To direct this new enterprise, Makukhin recruited a man of only thirty-five named Victor Brukhanov. With a reputation for great competence as an engineer, Brukhanov was both diligent and ambitious. After graduating from the polytechnic institute in Tashkent, where he had been born to Russian parents, he had studied hard to master the new technology of turbine engineering, and when he became head of a workshop, he had joined the Communist party. When Makukhin offered him the job at Chernobyl he was deputy chief engineer at the Slavanskaya nuclear power station.

It was a wonderful opportunity for such a young man. When Brukhanov first arrived in the town of Chernobyl, the whole enterprise consisted of nothing more than some plans in a briefcase and a rubber stamp. He took a room in the only hotel and then drove with Makukhin to the site where construction was due to begin. It was winter and there was snow up to their knees when Brukhanov drove the first wedge into the ground.

To begin with, Brukhanov had a hundred men and women working under him, and he had to find somewhere for them to live. He got hold of some rudimentary mobile homes—wooden huts on small metal wheels—and established a small settlement in a clearing in the woods. In August of 1970 he himself went to live there and was joined by his wife, Valentina, her mother and their two children.

Brukhanov soon realized that he had taken on awesome responsibilities. He had to supervise the construction of both the power station and the town to house the workers; and before either could be started he had to build sidings for the delivery of supplies and a plant to make the cement. Every month he had to certify expenditures, which started at 77,000 rubles per month and rose over the years to 120 million. Often he had to fiddle the accounts because the plans he was given were impractical. While waiting for the planned supermarket to be built, he found the money for smaller shops to sell groceries to the workers.

Goods came by rail to the sidings, and the enterprise could be fined if it did not release the rolling stock in a short space of time. Yet before the materials could be accepted their specifications had to be checked, and often they were discovered to be deficient. Caught between the demands of the planners on the one hand and the shortcomings of the suppliers on the other, Brukhanov found that he was expected to perform a superhuman task.

The initial plan for the power station was produced by a number of different institutes. The overall design, first drawn up by Elektroprojekt in the Urals, was later taken over by Zukh-Hydroprojekt in Moscow, whose expertise was principally in hydroelectric power. The plans for the reactor came from Dollezhal's bureau, NIKYET.

Although it was a tried and tested design, the simple expedient of increasing the reactor's output by increasing its size had led to an engineering project of gigantic dimensions. The reactor core was a huge graphite block weighing 1,700 tons. Like an immense honeycomb, it was penetrated by large-diameter machined holes: 1,661 channels for fuel assemblies—zirconium alloy tubes filled with pellets of uranium—and a further 211 channels for the boron control rods, which, when lowered into the core, absorbed the neutrons and either reduced the rate of fission or brought it to an end.

A plethora of piping brought water from six huge pumps into each of the fuel channels, where it was turned into a mixture of steam and scalding water by the heat generated by the nuclear fission. It then rose into drums, where the steam and water were separated, the steam going on to the turbines to generate electricity while the water returned to be recirculated through the reactor core.

Besides this principal circuit taking water through the reactor, there was a secondary circuit carrying water from reservoirs through the condensers and an emergency core-cooling system designed to protect the fuel assemblies should the main system fail or prove insufficient. This required piping and pumps of its own, and although it was considered an advantage of the RBMK's design that the rupture of a fuel assembly caused by overheating and consequent melting of the zirconium casing could be contained within a single channel, and that the water and steam were never under great pressure, it was still essential that every seam, every valve and every meter of piping should be flawless, and every installation entirely sound.

What Brukhanov discovered, however, was that the parts specified by the designers were frequently impossible to find. The industrial base existed to build the RBMK-1000s, but its productive capacity had not kept pace with the expansion of nuclear power. The Chernobyl nuclear power station had to compete with the other RBMKs being built at Ingalina and Kursk by the Ministry of Medium Machine Building. This huge, secretive institution had long-standing links with suppliers and could exert the kind of pressure that came from its contacts with the armed forces and the KGB. Retaining its monopoly in the mining and processing of uranium, upon which all nuclear power stations depended, it also controlled the production of gold and precious stones. This wealth put Slavsky's officials in a strong position when negotiating with other branches of industry or the planners at Gosplan.

Outbid in this way by his rivals, yet under great pressure to launch the first unit by 1975, Brukhanov was frequently obliged to manufacture the components he needed in workshops built on site. This encouraged a

spirit of improvization, which, though common enough in Soviet industry at the time, was dangerous when it came to nuclear power.

2

Besides this unfair competition with the Ministry of Medium Machine
Building, Brukhanov had to deal with the anomalies of the Soviet system
itself. First of all, there was the party, which acted as a shadow administration in every social, political, industrial or cultural structure. With few
exceptions—Dollezhal was one—anyone who wished to hold a position
of authority was expected to join the party, and, once a member, he
became answerable to those above him in the party organization. Thus
Brukhanov had to report not just to the officials of the Ministry of
Energy and Electrification but also to the regional party committee in
Kiev. It was the regional committee he feared most, because in the Soviet
Union the party had a monopoly of power. In theory one could move
from one job to another, or from the management of an industrial
enterprise into the civil service, but wherever one worked, one's pros-
pects depended upon one's standing in the party—and, of course,
clearance by the KGB.

Although some idealists were still to be found in the party hierarchy,
the majority of those who parroted Leninist slogans did so to further
their own careers. Besides the glory of holding high office, there were
more tangible benefits, like a higher salary, better housing, a car; reserved shopping facilities, holiday hotels and sanatoriums, passports and
hard currency for travel abroad, and access to the best educational
opportunities for their children. However, promotion came only with
performance, and for this the *aparatchiks* depended upon Brukhanov.

A small, curly-haired man with a mild manner, Brukhanov was ill-
equipped to stand up to the party bosses in Kiev. The kind of brutality
that had been found in the Communist officials who had administered
Stalin's terror in the 1930s had evolved forty years later into a bullying
manner, crude language and a threatening tone of voice. It was said that
you could always tell a party functionary because he had the face of a
truck driver but the hands of a pianist. "What is the first thing you want

to do when you reach the top?" asked Mikhail Zhvanetsky, the Jewish humourist from Odessa. "Spit down!"

To cover up their shortcomings, the party bosses took advantage of the dogma that the party could never err. In implementing the most recent Five-Year Plan, they were merely obeying the will of the party, as made manifest in the most recent Congress or a decree of the Central Committee. Its leaders, as the heirs to Marx and Lenin, were not only politically all-powerful, but morally infallible too. It was therefore impossible for Brukhanov to criticize the Plan. Since the party was always right, anyone who failed it was by definition either a criminal or a saboteur, and any criticism that suggested that there were shortcomings in socialism was counterrevolutionary.

Everyone knew that the party's rhetoric was false; they learned to live with lies. A large part of the population withdrew into themselves, with little expectations of life outside a small circle of family and friends. Already possessed of the Slav temperament—moody and sentimental, passionate yet apathetic, prone to hypochondria and self-dramatization, and with little love of work for its own sake—they ignored the collectivist slogans that were put out by the party and brandished on banners above public buildings. C. G. Feifer, an American correspondent in Russia at the time Brukhanov was building the power station at Chernobyl, remarked that no one he knew felt driven to get a job done, or even to go to work when not in the mood. The foreman could usually be persuaded to overlook an odd day's absence: a favor to be returned in due course with a bottle of vodka or a dozen eggs from one's mother-in-law's dacha. But even if one went to work, little was done. "On a given day, in any given office," Feifer was told, "eighty percent of the staff are in the corridor gossiping, going out to pee or comb their hair, or making a glass of tea."

Besides this general distaste for hard work, there were particular difficulties with the work force at Chernobyl. Gone were the days when red banners on the construction site or exhortatory Bolshevik slogans would inspire heroic achievements. Makeshift accommodations in the middle of nowhere had not attracted either skilled or experienced workers. The workers were mostly young, and the quality of their workmanship was poor. The head of construction, Vasili Kizima, was a tough and determined man, but neither threats nor exhortation could provide skills for the unskilled or persuade them to spend their evenings going to night school rather than getting drunk.

These problems with both labor and supplies meant that construction at Chernobyl fell behind schedule. After only a year as director, Brukhanov bitterly regretted that he had ever taken the job and tendered his resignation. It was refused. He remained as director and, leaving the problems of construction to Kizima, started to recruit the operating personnel.

3

The most important post to be filled was that of chief engineer. Because of the rapid expansion of the industry, there was competition for the best men, and Brukhanov's first choice left soon after his appointment to work for the Nuclear Safety Committee in Moscow. His successor, Akinfiev, came from the same military facility in Tomsk where Academician Legasov had worked in the 1960s. To head the turbine unit, Brukhanov hired a man named Taras Plochy, whom he knew from his days at the Slavanskaya power station. Nikolai Fomin, a Russian from the Donets region, was made head of the electrical workshop; like Plochy, his background was in conventional power generation, and there were whispers that Brukhanov chose his deputies from his own field because he felt intimidated by specialists in the nuclear sphere.

In the summer of 1973, during his summer vacation, a nuclear engineer named Anatoli Dyatlov came to Chernobyl prior to applying for a job at the nuclear power station at Kursk. On Akinfiev's recommendation, Brukhanov offered Dyatlov the post as deputy head of the reactor workshop. With many years experience installing small VVER reactors into nuclear submarines in the Soviet Far East, Dyatlov came with the highest recommendations: the management at Chernobyl was delighted to snatch him from under the eyes of their rivals at Kursk.

Besides his qualifications as a nuclear specialist, Dyatlov was politically sound—a good example to the younger workers of how a man from the humblest background could flourish under socialism. He was the son of a Siberian peasant whose job had been to light the buoys each night on the Yenisey River. At fourteen, when his father died and the village school closed for lack of students, he ran off to Norilsk, a city in the Arctic Circle that in winter hardly saw the light of day. There, after four years at a vocational school, he worked as an electrician, studying by night to qualify for further education. In time, he won a place at the prestigious Moscow Institute of Physical Engineering. It meant that his wife had to live with her parents in Vladimir: for six years Dyatlov lodged in a student dormitory and saw his family only on weekends. In 1959 he graduated and went back east, this time to Komsomolsk on the Amur.

This majestic river, which for a thousand miles forms the border between China and the USSR, flows into the sea of Okhotsk by the island of Sakhalin. Two hundred miles inland is the city of Komsomolsk, founded in 1932 as part of the first Five-Year Plan. Built by Komsomol volunteers, by the 1960s it had grown into a city of two hundred thousand inhabitants, with oil refineries, steel works and heavy engineering. It was here, in the greatest secrecy, that the small VVER reactors developed by the Kurchatov Institute were fitted to Soviet nuclear submarines.

In the naval shipyards, it was Dyatlov's task to assemble the active zones of these reactors and then test them both on shore and at sea. He liked the work. He was well-paid by Soviet standards; he could afford to send his family on holidays to Russia, the Caucasus or the Crimea, although he could rarely go with them because it was in summer that the submarines underwent trials in the Sea of Japan.

Dyatlov rose to be the head of the physics lab, leading a team of young nuclear engineers. He was a difficult man to work for—demanding, despotic and aloof. Having reached his present position by dint of his own efforts and innate intelligence, he was intolerant of the shortcomings of others. His knowledge was not limited to science; he loved literature and had collected the entire Library of World Literature that Maxim Gorky had started in the 1930s. But despite his arrogance and tactlessness, he inspired admiration as well as fear in the younger engineers. He was an outstanding nuclear specialist, and they were eager to learn from him. Some compared him to a snake that mesmerizes its victims; it would never occur to any of those who worked under him to question what he said.

Komsomolsk was not a pleasant place to live. It was devoted to its industries; the cultural amenities were poor. Almost everyone who worked there hoped to move on after a couple of years. For a time Dyatlov was the exception; only after more than ten years did he start to feel restless and consider looking for another job.

4

By the time Dyatlov started work at Chernobyl the construction of the first two units was well under way and plans had been drawn up for two more. "The bigger the better" had become an axiom of Soviet planners, and it applied not just to the enormous generating capacity of each unit but also to the number of them at any one location; by the mid-1980s there were six RBMK-1000 reactors either operative or under construction at Chernobyl.

The building that arose was huge and bland. A rectangular white block like a large shoe box housed the turbine halls, while the reactors were built in the square wings. Inside these geometric shapes was a mass of piping and machinery, but the exterior was clean and smooth. When the four units were completed, there were three chimneys—one, by the fourth unit, painted red and white—and outside the building there were clusters of pylons carrying the electricity to Kiev and beyond.

On either side of the power station were smaller buildings—stores and workshops of various kinds—and near the first unit was the administrative block where Brukhanov had his office. In the basement of this block was a series of interconnected rooms known as "the bunker": living quarters, a communications center and a clinic for use in the event of an emergency.

Seeing the power station for the first time was always extraordinary for those who came to work there because this temple to modern technology was set in a such a backward part of the world. On the road from Kiev, visitors would pass through a landscape of ponds, slow-flowing rivers and huge marshy meadows. Solitary peasant women, scarves tied tightly around their heads, tended herds of cows, as their ancestors had done for hundreds of years, while their husbands brought back their crops from the fields in horse-drawn carts. The only signs of the twentieth century were the occasional tractor or motorbike with a sidecar, and the long lines of pylons carrying power across the mournful landscape toward the pine forests to the north.

While the peasants who lived around Chernobyl spoke their own dialect and had lived there in ancient wooden cottages for many generations, the dazzling new town of Pripyat, which grew up only two kilometers from the gigantic power station, was inhabited by people from all over the Soviet Union. It was a young town—the average age of the inhabitants was around twenty-six—and exemplified the Soviet way of life. It was well planned, with well-spaced, eighteen-story blocks of flats. There were shopping centers, sports facilities, five schools, three different swimming pools and a permanent amusement park with a Ferris wheel. The city soviet, or town hall, was known as the "White House," and besides a hotel with two hundred beds, called the Polessia, there was a government guest house with 104 rooms and four suites for visiting party grandees.

There was no church, of course, but there was a cultural center for poetry readings and theatrical productions, as well as cinemas and excellent libraries. The town was well placed. To those who had lived in faraway places like Komsomolsk, Pripyat seemed to be at the heart of the nation. Kiev was only two and a half hours away by hydrofoil, and it was only a long day's drive to Moscow, Minsk, Moldavia or the Black Sea. To those who had waited on line for scarce groceries in some of the larger cities, Pripyat was a place of abundance. Thanks to energetic lobbying by Brukhanov, the shops were full of food; in one butcher's store, a newcomer counted fourteen different kinds of meat and sausage! And if an item could not be found in Pripyat, it was a simple matter to drive over the border into Belorussia, where food was always in good supply.

Better than the abundance of food or the facilities, however, was the proximity of the countryside. In a nation so recently industrialized, where most of the scientists and engineers were either the children or grandchildren of peasants, it was common to feel a strong bond with nature. In Pripyat, there was no pollution—nuclear fission gives off no fumes—and the town was surrounded by forest. A short walk or bicycle ride found the operators of the nation's most sophisticated technology among the birches and pine trees, fishing in the many rivers and lakes, collecting berries or mushrooms or letting their children run wild in the woods and bathe in the river.

The society there was also relatively egalitarian. The bosses had their perks, of course: Brukhanov was driven around in a white Volga and also had a dacha; but in Pripyat his flat on Lenin Street looked like all the others. There were rumors that it had been surreptitiously enlarged by the builders at the expense of his neighbors, and that a blue bathroom suite, intended for the Palace of Culture, had been appropriated for the director, but even if true, these were modest forms of privilege and corruption.

Indeed Brukhanov, who as director was responsible for the city, was much liked by the people of Pripyat. Not only was he free of the bullying manner so common in other leaders but he exerted every effort to make Pripyat a pleasant place to live. While in most other Soviet cities there was a waiting list of many years for housing, a family moving to Pripyat was soon allocated a flat; Brukhanov had calculated, quite accurately, that this was more important than any wage differential in attracting the best workers to the Chernobyl power station. He had also made sure that the shops were well stocked, and he had used spare supplies of steel to make greenhouses so that his people could have access to the un-heard-of luxury of fresh tomatoes. Because of his passion for roses, they had been planted in abundance in the municipal flower beds. Every aspect of life in the community was his responsibility; for example, when a fireman, Leonid Shavrey, wanted to buy a motorbike he had to apply to Brukhanov in person for permission. When a light bulb in a street lamp failed, it was Brukhanov who was informed and Brukhanov who had to get it changed. He worked so hard to live up to his responsibilities that the chief disadvantage of his solicitousness was not immediately apparent: by directing so much of his energy to ensuring a good life in Pripyat, he was distracted from running the power station.

5

In 1976, as the first unit was nearing completion, a group of young graduates from the Moscow Power Engineering Institute arrived at Chernobyl and went to live in the wooden huts of the forest settlement. With the dearth of experienced engineers like Dyatlov, many of the specialists came straight from the universities and polytechnics in this way: they were young men who had learned the theory behind the generation of nuclear power and certain basic skills, but who had also been taught a conformity and respect for authority found in all branches of the Soviet educational system. They had not been encouraged to show initiative or display a spirit of inquiry. Nor did they have access to training with simulators, which were available to young nuclear operatives in the West. They had to be trained on the job by the older engineers. Kizima, the head of construction, described Chernobyl as "the first university of atomic construction."

Some graduates were bright. Students who graduated at the top of their class could express a preference for where they wanted to work, and Chernobyl was often their first choice. They were from the most diverse backgrounds, reflecting the ethnic and national variety of the Soviet Union. Nikolai Steinberg, who came straight from the Moscow Institute of Electrical Engineering with a degree in hydraulic physics, was a Jew from Odessa; the first language of his grandfather, a banker, had been French. Razim Davletbayev and his wife, Inze, were both Tatars from Bawly in Kazan.

Following Kizima's dictum that they had to learn on the job, even graduates with the highest honors started at the bottom. Razim worked as a laborer and studied in the evening to qualify for promotion. Housed in the little wooden huts, the new workers would pore over plans laid out on the floor. It was little better than camping, with communal kitchens and communal showers, but they were all so glad to be involved in this prestigious project that no amount of hardship could restrain their zeal.

The older specialists, several of whom had followed Dyatlov from Komsomolsk, were as eager to teach the young as the young were to learn. There was no question of anyone expecting extra money for extra effort: Razim Davletbayev spent not only the evenings but also half his summer vacation studying the new technology of nuclear power. His wife, Inze—a woman with a wide Tatar face and gentle, soulful eyes— had a job in the department of industrial safety, checking the amount of radiation received by each worker, but all her ambitions were for her husband; she saw to all the household chores so that Razim could continue with his studies. After the birth of a baby, she spent much of the day with the child: Razim left for work at seven in the morning and only returned at nine at night.

Among Inze's closest friends was Luba, the wife of Alexander Akimov, whom Razim had known at the Institute in Moscow. Akimov had gone on to study at Zukh-Hydroprojekt, with the designers of the power station, and it was here that he had met Luba, also the daughter of an army officer and a student in the same department. Akimov had a gangling figure, thick glasses, a high forehead, receding hair and a small mustache. Luba was a tall, skinny girl with a delicate constitution, short dark hair and a sophisticated sense of humor. She loathed pretentiousness of a bourgeois kind, and was choosy about her friends.

Upon graduating, Akimov was sent to work for Zukh-Hydroprojekt in Chernobyl, and Luba went with him as his wife. They moved straight into a flat in Pripyat, where Luba gave birth to their first child. They, too, embarked upon the life at Chernobyl with the greatest enthusiasm. Akimov worked hard to establish his professional reputation; he also joined the party. In his free time, he read historical biographies, subscribed to magazines on military technology, and went after duck and hare with his Winchester rifle.

The Akimovs' life was not without trouble. Their second child was born with a twisted hip: every two weeks Luba had to make the five-hour journey on the hydrofoil to take her baby to see a specialist in Kiev.

6

On 26 September 1977, two years behind schedule, the first unit of the Chernobyl nuclear power station was finally commissioned. Despite the delays, it was hailed as a triumph. One thousand megawatts of electrical power could now be fed into the grid at Kiev, a saving in terms of thermal power of three and a half million tons of coal each year. Brukhanov and Kizima were praised and decorated. Soviet science and technology had triumphed yet again.

Razim Davletbayev, who had worked as a simple laborer at the start-up, was promoted to turbine engineer. Alexander Akimov, now a specialist in the automation of electrical power, was transferred from Hydroprojekt onto the staff of the power station itself. He served under Nikolai Steinberg, who, with the departure of Taras Plochy to be chief engineer at the Balakovsky nuclear power station, was now chief of the turbine hall. It was a particularly cheerful unit; Steinberg, who had a keen eye for members of the opposite sex, had taken on an exceptionally pretty girl called Katya Litovsky, the only woman among twenty-two men. This did little for her marriage—she soon separated from her husband—but much for the morale of the engineers.

Akimov, a more earnest character than Steinberg, served as Communist party secretary for the unit—a chore that reflected his commitment to communism and also helped his career. His friend Razim Davletbayev also joined the party, although Inze tried to dissuade him; she thought there were as many good people outside the party as in it and resented the way in which her mother, who had toiled for twenty years, had never been invited to join until she had taken a degree.

At a higher level, Nikolai Fomin, who had been head of the electrical workshop, laid aside his professional duties to serve as party secretary for the power station as a whole. Under the system it was a position of both power and responsibility; as the elite, party members were answerable for the successes or failures of their organization; and, whatever their rank in the station's administration, they had to defend themselves against criticism at party meetings.

Fomin, although already middle-aged, also studied physics by correspondence to add to his existing qualifications. When he returned to nonpolitical work he was rewarded with the post of deputy chief engineer.

Little more than a year after the launch of the first unit, on 21 December 1978, the second unit went on line. Knowing only what they were told by their superiors or had read in the Soviet press about the absolute safety of Soviet reactors, none of the operators realized that

three years before there had been a meltdown of a fuel element in the
No. 1 unit at Leningrad, which was identical in design to the unit at
Chernobyl. Such was the secrecy that still surrounded anything to do
with nuclear power that the accident was concealed even from those who
worked in the industry. There was little opportunity to learn from the
mistakes of others.

Nor was this accident at Leningrad the first; there had been others at
Leningrad and at Beloyarsk. Only ten days after the commissioning of
the second unit at Chernobyl a fire at Beloyarsk burned through the
control cable and the reactor momentarily went out of control. No news
of these accidents reached the outside world, and the Chernobyl operators proceeded on the assumption that if they kept to the regulations in
the documentation that went with the reactor—drawn up by a department of the Ministry of Energy in the 1960s upon the advice of the
scientific supervisor, Anatoli Alexandrov, and approved by the State
Committee on Nuclear Safety and the chief engineer of the Chernobyl
nuclear power station—there was no possible danger of a serious accident.

There were certain minor contradictions, but this was nothing new.
Modifications made to the later reactors suggested that the earlier ones
were not perfect; for example, in the first two reactors the control rods
were lowered into the reactors by manual operation of the drive motor
switches, and it was only later that automatic controls were introduced.
It was also generally understood that with so much piping, there was
always the possibility of a rupture of some kind, but the possibility of a
major accident never entered anyone's mind. An accident was conceiv-
able in a pressurized-water reactor, with its greater power density and
higher fuel rating, but not in the RBMKs.

If the RBMKs were safe, however, they were not always easy to
control. It was an advantage that they did not have to be closed down
for refueling; one fuel element could be replaced at a time. But the
number of dials that the operators had to monitor at any one time, as
well as the the complexity of the information from the Skala computer,
which scanned the reactor's condition only every five minutes and
printed out its findings fifty meters from where the operators were
stationed, made it difficult for them to understand at all times exactly
what was going on.

The operators also discovered that the reactor was unstable when run at low power: in the same way that a jet aircraft might stall at low throttle and crash, power could collapse in the reactor. It therefore became accepted practice, when it looked as though the reactor would die on them, to withdraw more of the boron control rods than was permitted by the regulations. The regulations governing safety were not set in stone but could be overruled upon the authority of the chief or deputy chief engineer. Over the years of running the RBMKs, the engineers had decided that the parameters recommended by the designers bore as much relationship to the reality of running the reactors as did the Communist rhetoric to the realities of life in the USSR.

There were also risks to be run by playing safe. An emergency shut-down of the reactor—easily done by pushing the AZ button, which automatically lowered all the rods into the core—led to a catastrophic cut in the generating capacity of the nuclear power station and in loss of electricity to the grid. It was estimated that an unscheduled shutdown would cost the power station six hundred thousand rubles in lost revenue. If it was subsequently considered unnecessary, it could lead to demotion, loss of salary, even dismissal and expulsion from the paradise of Pripyat. A shutdown was therefore not something a young operator would want to do on his own authority; he would always prefer to seek the approval of a senior engineer.

When news filtered through about the accident on Three Mile Island, it only seemed to confirm the truth of what the operators at Chernobyl had always been told: that where profit was at stake, the capitalists would take insensate risks with the safety of the local population. Molded by their Communist education, taught to accept without question what they were told by their superiors, isolated by an all-pervasive censorship from any hint of doubt, they never questioned the assurances given by the eminent physicists in Moscow. If the president of the Academy of Sciences himself, Anatoli Alexandrov, insisted that Soviet reactors were entirely safe, who were they to doubt him?

They were aware, of course, that minor accidents could occur—hence the leak-tight containment and contingency plans for a ruptured seam or a burst pipe. They also knew that there might be a dangerous release of radioactivity in the immediate vicinity of the power station—hence the bunker under the administrative building—but they all considered the likelihood remote. When given the task of drawing up a program for such a hypothetical accident, Alexander Akimov calculated a probability factor of one in ten million per year.

The most likely moment for such an accident was during the start-up of a reactor, when all the circuits were being used for the first time. In December 1981 the third reactor was successfully launched, and the new input of one thousand megawatts into the grid enabled the first reactor, which had been running for five years, to be closed down for routine maintenance in the summer of 1982. In the course of the first reactor's restart in September the valves controlling the flow of water into the reactor were inadvertently closed. Some of the fuel assemblies over-heated, the uranium melted, and there was an explosion in the core and a release of radioactivity into the power station, some of which escaped through the filters into the air outside.

No one was killed. The emergency core-cooling system did its work, and the reactor was closed down. Engineers repairing the damage received significant doses of radiation. Outside the plant, no measurements were taken. The streets of Pripyat were hosed down by water tankers as a precautionary measure, but no one was told of the accident; indeed, so thorough was the regime run by the KGB's department of classified information that even the operators in the other two reactors did not know what had happened in unit No. 1.

From the point of view of the management and the Ministry of Energy, the most serious consequence of the accident was not the minor release of radioactivity but the eight months it took to repair the reactor. This meant a large loss of revenue to the enterprise, and as a result heads had to roll. The chief of shift and a deputy chief engineer were both demoted, and the latter was transferred to the thermal power station at Zaporozhye. But the buck did not stop there. After an investigation by the Kurchatov Institute and Dollezhal's bureau, the Party Committee in Kiev decided that the chief engineer, Vyachslav Akinfiev, should be held responsible. He was demoted to the level of deputy chief engineer and removed from any responsibility for the running of the plant.

For Akinfiev this was intolerable. He did not feel that the accident had been his fault; he thought that the deputy chief engineer alone was to blame, and, having been a leader of the enterprise since its inception, he felt humiliated to be reduced to a subordinate role. Although no one dared ask what his crime had been, his fellow workers knew that he was now in disgrace: he was cold-shouldered by the very colleagues who once had fawned on him. Therefore he resigned and, through contacts in Moscow, was offered a job at a new VVER power station then being built by Soviet engineers at Kozloduy in Bulgaria.

7

Akinfiev's demotion created a vacancy for the post of chief engineer. Among the candidates was Nikolai Fomin, then a deputy chief engineer. A competent and hard-working electrical engineer who knew no more about nuclear physics than he had learned from his correspondence course, he could now draw on the ideological credit he had banked when serving as party secretary for the plant. It was enough, it was argued, to have atomic specialists like Anatoli Dyatlov and Mikhail Lyutov as deputy chief engineers. For the post of chief engineer it was more important that the man should be an effective leader. Fomin's irascible, dominating manner was thought, like Kizima's, to counterbalance Brukhanov's gentler style.

Fomin got the job, whereupon the intrigue now centered on his replacement as deputy chief engineer for units 3 and 4. The choice narrowed down to two—Steinberg, aged thirty-five, or Dyatlov, fifty-two. Brukhanov himself favored Steinberg, but Dyatlov could call on contacts he had made during his years in Komsomolsk. Moreover Steinberg was a Jew, and while Jews, like the Volga Germans, were often the most able engineers, they were also often the victims of envy and prejudice among Slavs in the Moscow ministries and the party apparatus in Kiev.

Dyatlov was appointed. To the team of engineers who had followed him from Komsomolsk—men like Sitnikov, Chugunov and Grishenka—the ministry had chosen the best man for the job. Fomin was an electrical specialist; Steinberg a turbine engineer. It was essential to have a nuclear specialist in the upper echelons of the plant's direction, and Dyatlov was, in their opinion, by far the best physicist on hand.

There was also an element of self-interest. The Soviet educational system turned out a large number of engineers, and since there could be no overt unemployment in a socialist state, these new graduates were allocated to different enterprises throughout the country. As a result, Chernobyl, like any other power station, was obliged to employ many more technicians and specialists than it actually required. Whereas each shift at a nuclear power station in Western Europe or the United States might consist of ten or fifteen operators and engineers, at Chernobyl there would be seventy. Many, naturally, would have nothing to do, and bored workers were sometimes found reading novels or playing cards.

The management dealt with this overmanning by creating an inner elite to run the station, called the Group of Effective Control. Everyone started at the bottom and worked for a time in the various departments, but through a process of natural selection the ablest were put into teams that actually ran the station.

As in any organization, however, there was more room at the bottom than at the top, so even for these high flyers there was a limited chance of promotion. With Fomin in place as chief engineer; Dyatlov, his deputy, in charge of units 3 and 4; and Dyatlov's friend Sitnikov overseeing units 1 and 2, patronage for the foreseeable future was firmly in the hands of the group from Komsomolsk. Steinberg therefore decided that the time had come to move on, and with the highest references from Brukhanov he went to the Balakovsky nuclear power station, where his old friend Taras Plochy was chief engineer.

When Steinberg left Chernobyl, about thirty other specialists went with him, realizing that they could expect no promotion under the new regime. This demoralized many of those who remained behind—not just the pretty Katya Litovsky, whom Steinberg had brought into the turbine unit, but some of the more open-minded engineers who disliked the dogmatic and dictatorial style of both Fomin and Dyatlov.

Of the two, Fomin, with his deep voice and pleasant smile, had the more agreeable manner. He could, however, be brusque, and he disliked criticism. "We're not at a party meeting here," he would say to any subordinate who dared to question his instructions. In the opinion of some of the nuclear specialists from Komsomolsk, he was overconfident about his own abilities in their sphere of expertise. He certainly had no doubts whatsoever about the safety of the reactors. He often reassured visitors by comparing the reactors to samovars; when the Ukrainian minister of health, Anatoli Romanenko, visited Chernobyl Fomin told him that the chances of an accident there were much the same as being hit by a comet. "How many comets hit the earth? One almost every hour. Yet we know of only two cases where they caused fatalities: one hit a man going for a walk, and the other fell into the washtub where a German woman was doing her laundry. No, Comrade Minister, the chances of a serious accident with a reactor are about the same as being hit by a comet."

Fomin spoke in all sincerity, for though he knew that there had been an accident at the recommissioning of the first unit, he also knew that the emergency systems had done the job for which they were designed. Certainly there were minor accidents, as there were in any industry—on building sites, at steel works and in power stations burning coal. But even the disaster at Three Mile Island, where the profit-hungry capitalists had been indifferent to the fate of the workers, had injured no one, and it was only as a result of the sensational reporting of the accident in the irresponsible capitalist press that people living in the vicinity had panicked and fled from their homes.

Dyatlov was a more complex character. A tall, wiry man with high cheekbones and a receding hairline, his quizzical expression made him resemble the celebrated bust of Voltaire. An unhappy childhood spent near the penal settlements in the Krasnoyarsk region in Siberia seemed to have left him with a gloomy outlook on life. As with many self-made men, he was impatient with the failings of others. He was also tormented by the conflicts inherent in the system. On the one hand he was an active party member, carrying the banner of his section at the May Day parades in Pripyat, and he was grateful to a social system that had enabled him to rise from a poverty-stricken background in the remotest corner of Asia to his position as a prominent and powerful engineer.

Yet Dyatlov could not hide from himself the deficiencies of the system, or fail to notice the shroud of falsehood that concealed them from the outside world. This was particularly evident in the work he was engaged in, the construction of the fourth reactor. The plans were inadequate and called for materials that were either unobtainable or were delivered late and were frequently defective when they finally did arrive. The fireproof roofing and electrical cable specified in the plan were not to be found, and so the builders had to improvise, and the supervisors to turn a blind eye to violations of the regulations. In the middle of the unit there was a "dead zone" where nothing could be built because the plans had not arrived; yet construction continued around it in the hope that the two different sections would eventually fit together.

This chasm between rhetoric and reality was like a painful fissure in Dyatlov's own personality. He had done everything the system expected him to do; he had read every book in his Library of World Literature and knew Pushkin's *Eugene Onegin* by heart; yet neither the science nor the culture that he had acquired so diligently had made him a happy man. As a result he was irascible and domineering, and the younger operators feared him. Even his friends from Komsomolsk, whom he saw on his time off, were often uncomfortable in his presence. He inspired their respect but not their affection. Only Anatoli Sitnikov, who had started with him all those years ago installing reactors in nuclear submarines, could talk to him on an equal footing. His wife adored him—or so it seemed to their friends—but she was a small, cozy woman, not the type to tame the tyrannical traits in her husband, and there was the shadow of a shared sorrow between them: their second child had died in infancy in Komsomolsk.

Dyatlov was also getting older, and many men in their middle fifties begin to feel the strain of a demanding profession. He delegated certain duties, such as the training of personnel, and during the laborious months leading to the start-up of reactor No. 4, many of the operators felt that he gave them no real guidance but left it to the section leaders to settle matters between themselves.

8

As the autumn of 1983 turned to winter, pressure built on the engineers at Chernobyl to complete reactor No. 4 by the end of the year. To do so would mean that they were a year ahead of schedule and would be eligible for numerous bonuses and awards—not just for the workers, but for everyone involved, including Dyatlov, Fomin, Brukhanov, and those above Brukhanov in the Ministry of Energy and Electrification.

As the year drew to a close, what held the engineers back was not the construction but the numerous tests that had to be carried out by the government commission responsible for certifying the safety of the plant. In particular there was the vexing question of how to deal with an unexpected cut in the supply of electricity to the control-rod drive mechanism, and both the main circulating pumps and the emergency core-cooling system. Earlier tests had not had satisfactory outcomes.

Unknown to the operators, there had been two further accidents since the destruction of the fuel assembly in reactor No. 1 in September 1982. Only a month later there had been an explosion in a generator at the Armyansk nuclear power station, after which the turbine hall had burned to the ground; and in June 1985 there had been a horrific accident at the pressurized-water nuclear power station at Balakovsky, to which Taras Plochy and Nikolai Steinberg had gone from Chernobyl. Because of mistakes made by operators under Plochy's command, during the recommissioning of reactor No. 1, a valve had burst and sent steam at 300°C (572°F) into the area around the well of the reactor, where fourteen men were working. All were poached alive.

It was not this tragedy, however, that called for the tests on the turbines at Chernobyl, but rather an earlier and potentially more disastrous power cut at the Kursk nuclear power station in 1980. In the RBMK reactors, electricity was required to drive both the control rods and the water pumps. In the event of an unexpected power cut, certain standby measures could be taken: the control rods could be disconnected from the power drive and allowed to drop into the reactor under their own weight, and also standby diesel generators had been installed that could provide an alternative source of power for the pumps.

The danger occurred during the forty seconds between the power cut and the start-up of the generators. At Kursk, in 1980, the natural circulation of the water had proved sufficient to prevent an uncontrolled power surge, but it was impossible to count on this. The turbine manufacturers hoped to be able to modify the generators to squeeze sufficient power for those few crucial seconds from the turbines themselves while they were spinning to a halt. However, it was clear that it would be impossible to conduct tests to verify the efficacity of this modification if the reactor was to be commissioned before the end of the year, and since the likelihood of a power cut was remote and the outcome at Kursk had been satisfactory, the tests on the turbines were postponed.

There was a further anomaly that had first been noticed by scientists from the Ministry of Medium Machine Building during the physical start-up of the first unit of the Ignalina nuclear power station that year. As the control rods were inserted into the reactor core, instead of an immediate decrease in reactivity, there was a momentary surge before the decrease began. This seemed to be caused by a flaw in the design of the control rods, and although it was also noted at the start-up of the fourth unit at Chernobyl—again by scientists from the Ministry of Medium Machine Building, who inspected the finished reactor—it was not thought sufficiently hazardous to be mentioned to the operators or written into the documentation.

On 21 December 1983, the fourth reactor was commissioned, and on 31 December, at the eleventh hour, Brukhanov signed a receipt for the completed reactor. The SPC certificate was signed by an official from the Ministry of Energy and Electrification. The bonuses were secure and the awards followed: for Brukhanov a Hero of Socialist Labor, for Fomin a medal for Valorous Labor, for Dyatlov an Order of the Red Banner.

On 27 March of the following year, three months ahead of schedule, and amid a fanfare of congratulatory publicity in the press, the fourth unit at Chernobyl went into commercial operation. For the opening, potted palm trees were laid out in the turbine hall, and not only the minister of energy and electrification and all the party leaders from Kiev, but also the president of the Academy of Sciences, Anatoli Alexandrov himself, came to Pripyat. He could look upon what he saw with some satisfaction. Four units, all powered by RBMK reactors, were now generating four thousand megawatts of electricity for homes and industries throughout the western republics of the Soviet Union, and even for the grid of the neighboring fraternal socialist republics in Eastern Europe. It was a triumph for Soviet science, for Soviet technology, and a significant contribution to the construction of the new Socialist civilization. While the Americans were faltering over nuclear power after Three Mile Island, the Soviets forged ahead to establish an undisputed ascendancy in the field. In the middle distance, Alexandrov could see the foundations of units 5 and 6, then under construction. When they were built and went on line, the V. I. Lenin Nuclear Power Station at Chernobyl would be the largest power station in the world.

III

1

If the shortcomings of the Soviet system were apparent to Anatoli Dyatlov, they were also evident to many political leaders. The rule of Leonid Brezhnev and his satraps in the provinces, like Slyunkov in Belorussia and Scherbitsky in the Ukraine, with their corruption and incompetence, had led to a stagnant economy. The insatiable defense industries continued to take an exorbitant share of the nation's resources. If there were shortages in the nuclear construction industry, there was also a dearth of consumer goods, and while censorship and an omnipresent KGB might suppress any overt expression of discontent, there was a malaise in the work force and an inefficiency in administration that no amount of propaganda could conceal.

In the last years of the Brezhnev era, and during the brief tenure of General Secretary Yuri Andropov, some of the elder statesmen—notably Andrei Gromyko, Mikhail Suslov and Andropov himself—recognized the need to promote younger, more energetic and open-minded leaders, and brought into the Central Committee the party boss of the Stavropol region, Mikhail Gorbachev. When Andropov died, the old guard on the Central Committee chose one of Brezhnev's old cronies, Konstantin Chernenko, as general secretary, but his period in office was brief, and when he died in March 1985, Mikhail Gorbachev was chosen to replace him.

Gorbachev immediately resumed the program of reform that had been started under Andropov. Assisted by like-minded allies on the Central Committee, such as Yegor Ligachev and Nikolai Ryzhkov, he

sought to revive the Soviet economy by replacing corrupt and complacent officials. Henceforth no official or institution was to be regarded as
beyond criticism, nor were party members to be immune from prosecution. Under the slogan *"glasnost i perestroika"*—openness and restructuring—these policies were adopted at the Twenty-seventh Congress of
the Communist Party in February 1986.

It was one thing to endorse reform and another to achieve it.
Throughout the Soviet Union, the vast party apparatus had a vested
interest in the status quo. Gorbachev could remove sixteen of the Soviet
Union's sixty-four ministers and 20 percent of local party officials by the
end of his first year in power, but at the grass-roots level his reforms
depended upon local initiatives, and it still took a brave man to criticize
a party boss. The style of leadership established by the Bolsheviks, which
mixed threats with Marxist slogans, was still the norm.

At Chernobyl, the first sign of *glasnost* was an article in the journal of
the Ukrainian Writers' Union, *Literaturnaya Ukraina*. It was not
unusual for a literary magazine of this kind to show an interest in
industry. In the Soviet state, writers played a special role; as if recognizing that socialism itself came from the fertile minds of intellectuals, the
expression of ideas was strictly controlled. Writers whose work met with
approval were admitted to the Union and given the kind of privileges
that came with membership in the Academy of Sciences; the work of
writers who were critical of communism was not published, and if
printed surreptitiously would count as evidence of criminal anti-Soviet
propaganda. Like a covey of poet laureates, the pampered members of
the Writers' Union were expected to play their part in building a Socialist
state. The proletariat had no patience with art for art's sake. Thus,
Literaturnaya Ukraina had for some years shown a lively interest in
industry, with a regular column entitled "Eye on the Chernobyl Power
Station."

Before the Twenty-seventh Party Congress, this eye had seen only
laudable achievements by heroic workers and engineers, but with the
advent of *glasnost* and anxiety in the highest quarters about the shortcomings in construction, it was decided to invite a journalist on the spot
to take a critical look at the empire of Comrade Kizima, the head of
construction at Chernobyl.

It was a measure of the way Pripyat had developed into a flourishing community that it now had its own newspaper, *Tribuna Energetica*, edited by a young woman named Lubov Kovalevskaya. She had first come to Pripyat as the wife of a lathe operator at the power station, and had worked as a schoolteacher while writing poetry in her spare time. In 1981 she was given a job on the staff of *Tribuna Energetica* and became a probationary member of the Communist party, which she joined in 1983.

Lubov had been born among the prison camps on the Black River near Sverdlovsk, the granddaughter of a Polish landowner who had been deported from Belorussia in the late 1920s as an enemy of the people. Her maternal grandfather had been shot because a passerby had heard him make a derogatory remark about Stalin; her grandmother, a kulak, had been sent to Siberia for hiding grain in the pocket of her dress. The flat where Lubov had grown up was in a wooden block built by the convicts of the prison where her father worked as an accountant. She had been something of a tomboy, but with beautiful gray eyes and a trusting, affectionate nature. She had met her husband, Sergei, while she was training as a teacher at Nizni Novgorod and, after they were married, went with him to Chernobyl.

In 1983 Sergei left the power station to take seasonal work on construction sites in the Arctic Circle. He spent the summer in the north and the winter in Pripyat. Meanwhile, Lubov's career flourished on *Tribuna Energetica*. She became the paper's acting editor and published poems about unhappy love. When foreign delegations came to visit the power station, she was invited by the management to entertain them at receptions, where she met such luminaries as Brukhanov, Kizima and Fomin.

When Lubov was asked by the editor of *Literaturnaya Ukraina* to write articles on the station, she found that many of those who knew of the shortcomings of the nuclear power station would only discuss them off the record. No one wanted their criticisms to appear under their own name. However, all the material she needed was close at hand. Only a few minutes' walk from the office of *Tribuna Energetica* there was the information center for the power station. Here all the records were stored on a computer. A friend of Lubov's who worked there could call up all the faults and delays that had taken place over the years. The pattern Lubov established was clear; the quality of the construction had deteriorated as time went on, reaching its lowest point with the fifth and sixth reactors now being built. The first four reactors were now producing four thousand megawatts of electricity, and the initial successes had been rewarded with various government awards, but because of these achievements "the lagging behind in the construction of the fifth power set is particularly noticeable. The plans and specifications for building and assembly work in 1985 have not been fulfilled. Is this slump a coincidence? Of course not. But it is not enough to give us a simple answer either."

In her article, Lubov went on to describe what had been apparent for many years but had never been expressed in public: the "late release and design and costing documentation" made by Hydroprojekt made it impossible to arrange the timely delivery of supplies, so the structures due in the summer arrived in the winter, when the weather made it impossible to install them. Lubov went on:

> Lack of organization weakened not only discipline, but also the responsibility of each and every one for the result of the work. The impossibility, or even unwillingness, of technical engineering workers to organize the teams' work lowered expectations. . . .
>
> There is yet another problem, a considerably more serious one. Due to the unscrupulousness of the supplier, the customer cannot rid himself of defective material and of disruption in supply. Thus, in 1985, 45,500 cubic meters of prefabricated reinforced concrete was ordered, 3,200 meters was missing, and of the 42,300 received 6,000 was faulty. Thus the reinforced concrete is there but it cannot be assembled. . . .
>
> I do not wish to suggest that the producer factories do not have problems of their own and complications, nor do I want to deliver any lectures. All the same, I consider it to be an abnormal situation when contract obligations are constantly violated."

Lubov then proceeded to name the organizations that had undersupplied sections for the cooling towers, wall panels for the machine hall, concrete slabs and pumping channels. She wrote that 326 tons of fissure sealant for the nuclear-fuel waste depository were defective, as were girders for the machine hall. Such faults, she argued, were inadmissible, above all in a nuclear power station,

> where every cubic meter of reinforced concrete must be a guarantee of reliability, and thus of safety. And the main factor governing everyone involved in the construction of power projects must be, above all, his conscience. I am convinced that anyone with a conscience finds defects unacceptable, because he finds them degrading.
>
> It is offensive and insulting to the worker to have to correct mistakes made by others because this shows, above all, a lack of respect for him. An absurd situation arises; the structure has arrived, but it cannot be used; yet there will be no replacement. Just how much self-control, sharp-wittedness, strength and nervous energy is required from the builder to finally put a structure like this in order?

Lubov wrote the article in a single evening. She would have liked to have left out the fawning references to the odious general secretary of the Communist Party of the Ukraine, Vladimir Shcherbitsky, or the directives of the Twenty-seventh Congress and speeches by the minister of energy and electrification on the Five-Year Plan. There were also things she would have liked to have put in: her contempt for the corruption she saw all around her in Pripyat; the way leaders gave well-paid jobs to their unqualified wives so that they could go for holidays to the Black Sea; the way senior managers took on their old friends in their departments and covered up for their errors; the way everyone spouted the slogans of collectivism while looking out only for himself.

On 27 March Lubov's article appeared in *Literaturnaya Ukraina* under the title "The Resolutions of the 27th CPSU Congress Being Put into Action. It Is No Private Matter." The immediate reaction in Chernobyl was one of anger and dismay. Although Lubov's criticism was chiefly directed at suppliers, it was thought disloyal to bring matters of this kind into the open. Brukhanov, who was always at odds with Kizima, was happy enough to see his rival discomfitted, while Kizima told Lubov that she was not qualified to write an article of this kind. It was said that she had sold out her friends in Pripyat to make a name for herself in Kiev, and rumors reached her that the town committee was considering expeling her from the party.

2

Two days before the publication of Lubov Kovalevskaya's article in *Literaturnaya Ukraina*, Nikolai Fomin, the chief engineer of the Chernobyl nuclear power station, had returned to work after a long period convalescing from a severe spinal injury. He had left his flat one evening without his keys, and when he got back he had tried to climb in through the balcony but had slipped and fallen onto the pavement. He had remained semiparalyzed in the hospital in Pripyat for a month and a half, but had returned to work as soon as the doctors would allow him. Grigori Medvedev, a former deputy chief engineer, visiting Chernobyl at the time in his capacity as an official of Hydroprojekt, was taken aback by the change he saw in Fomin. Medvedev suggested further convalescence, but Fomin said that there was too much to do.

If Fomin was an invalid, Brukhanov was under pressure from different directions. It was not just the construction of the new units that were plagued with difficulties but also the working reactors themselves. There were leaks in the air vents and drainage channels; up to fifty cubic meters of radioactive water escaped every hour and the extraction units could barely deal with it.

Brukhanov was constantly pressured by those above him in Moscow and Kiev. The Ministry of Energy in Moscow had ordered him to replace the inflammable roof of the turbine halls without telling him where he was to find a fireproof roof one kilometer long and fifty meters wide. He would have to petition for exemption. They had once complained that the cables did not have fireproof covers as the specification required, but had later agreed with Brukhanov that such covers were also unobtainable and had agreed to make an exception to the rules.

Quite apart from the problems facing the power station, there was a new directive from the government that every industrial enterprise had to diversify into some kind of consumer goods. Brukhanov knew the problems facing the economy: although only trained as an engineer, he had been a delegate at the Twenty-seventh Party Congress and had therefore voted along with all the others for the new party line. But what could Chernobyl produce other than electricity? He was told that his colleagues at Beloyarsk were already manufacturing souvenirs: an aluminum plaque showing a detachment of cavalry against a background of the steppes, something to appeal, no doubt, to their minister, Slavsky, who had ridden with Budenny during the civil war. But what could Chernobyl produce? The only suggestion was nuclear-powered meat-mincing machines!

Brukhanov was tempted to put his foot down and say, "A nuclear power station is not a craft shop. We will stick to what we know best and only generate electricity!" But it was not in his nature to put his foot down, and if he did so he felt sure that he would be either sacked or demoted, leaving his successor with exactly the same insoluble problems. Therefore he simply did his best to deal with each problem as it arose—the latest being a demand from the regional Party Committee in Kiev that the nuclear power station manufacture two hay-storage facilities for a neighboring collective farm.

With Brukhanov distracted and Fomin convalescent, the burden of the everyday operation of the power station fell on the shoulders of the deputy chief engineers. That April the load was particularly heavy for Anatoli Dyatlov, the irritable Siberian from Komsomolsk who, as deputy chief engineer responsible for units 3 and 4, had to supervise the shutdown of the fourth reactor for routine maintenance and repairs.

In itself, this was not a complicated procedure, but it was important to take advantage of the process to conduct a variety of tests on the equipment. Among them was to be a further attempt to measure the modifications made to the turbogenerators, which should have been made before the reactor went on line. It had still to be established whether, in the event of a power cut, the declining momentum of the turbines could generate enough electricity to power the pumps for the forty or fifty seconds before the standby diesel generators took over.

The test was to be conducted by the enterprise Donenergo, which had manufactured the generators. Their representative, Gennady Metlenko, had been sent to Chernobyl with a special mobile measuring unit, manufactured in the west by Mercedes-Benz, and borrowed from the Karkhov turbine factory. He had drawn up a draft program for the test; all that was required of Dyatlov was that the reactor be run on low power and then switched off so that the turbines could be disconnected and idle to a halt. Dyatlov discussed the proposal with Metlenko and with the heads of different workshops, like Piotr Palamarchuk, a tall, broad-shouldered Ukrainian from Vinnitsa. As the head of the start-up workshop, Palamarchuk had a team of twelve men conducting six different tests on the pumps, valves and turbines. Alexander Kovalenko, the head of the reactor workshop, looked over the proposal for a quarter of an hour and then approved it.

When the draft was agreed to, Dyatlov had it typed out and sent to Fomin for approval. Fomin, of course, was not a nuclear specialist, and Lyutov, the scientific deputy chief engineer with overall responsibility for nuclear matters, was away having a medical checkup. It was not thought necessary to consult Yuri Laushkin, the resident safety inspector, or to show the new program to the head of the shift, Boris Rogozhkin. The provisions for safety on the proposal were included in a purely formal way, but that was not unusual; they had remained more or less the same since they had first been drawn up in the late 1960s, and Dyatlov had not thought it necessary to rewrite each clause. Moreover, most of the regulations contained the proviso that they could be overridden by the chief or deputy chief engineer, and Dyatlov intended to be in the control room during the shutdown. Fomin approved the proposal and sent it back to Dyatlov.

3

In Pripyat preparations were under way for a shutdown of a different kind. May first was the major festival of the Communist year, and although it would fall on the Thursday of the following week, many meant to take time off and start their holiday on the previous weekend. Some planned to go away to relatives in the country; others would stay for the parade and the festivities in Pripyat, but even those who remained had to stock up on food and drink.

The weather was exceptionally fine. In the Ukraine, it was usual for the long winters to give way rapidly to spring, but it was so warm this May that it seemed more like early summer. The grass was green and the cherry trees were in bloom. The air was scented by the pine trees of the forests surrounding Pripyat, and everyone, from those who worked in the power station to the doctors in the hospital and the teachers in the schools—and above all the children sitting restlessly at their lessons—longed for the holiday to begin.

The different teams that ran the fourth unit worked in eight-hour shifts—three days on, three days off, two days on, two days off. At midnight on the night of 24 April, Alexander Akimov clocked in as shift foreman along with the twenty-six-year-old engineer Leonid Toptunov as his senior reactor-control engineer. Piotr Stolyarchuk was the senior unit-control engineer and Igor Kirschenbaum, only two years older than Toptunov, was senior turbine-control engineer. They were alert and eager to learn from the experience of the shutdown; to study what happened afterward, equipment had been installed to record what was said.

At 1:00 A.M., on Dyatlov's instructions, Akimov started to reduce the reactor's power. This could not be done quickly because the xenon gas that came from the radioactive decay of uranium 135 absorbed neutrons. If not allowed to decay, which it did in a few hours, it could accelerate the decline in the reactor's activity and close it down altogether.

At 8:00 the next morning, a new shift took over. At 1:00 P.M. the reactor reached half power and turbine generator No. 7 was switched off. At 2:00 P.M. everything was ready to start the test on turbine generator No. 8, but there was a danger that either the decrease in water supply to the reactor or the start-up of the auxiliary diesel generators would signal an accident to the sensors and start the emergency core-cooling system, which would then flood the reactor and close it down.

Since disconnecting the emergency system without the permission of the chief or a deputy chief engineer, was forbidden by the regulations, Dyatlov's authorization was required for Metlenko's experiment to continue. It was given, and once the system was disconnected, the operators were ready to reduce the power still further, to the point where the second turbine could be switched off, while keeping the reactor on low power for a short period in case the test had to be repeated; but at this moment a call came through from the load dispatcher in Kiev saying that the power generated by turbine No. 8 would be required by the grid until 11:00 P.M. This meant that the test had to be postponed.

The group of observers dispersed. Leaving instructions to the shift foreman to keep the reactor working at half power, Dyatlov now went home for some sleep. He was tired and frustrated by the delay. At 4:00 P.M. a new shift under Yuri Tregub took over in the control room of the fourth unit. Since it had been assumed that the experiment would have been completed by now, some were not familiar with the program and studied it for the first time. Seeing a section that had been crossed out, one of the operators telephoned another and asked if he should comply with the deleted section or not. His friend hesitated and then said, "Yes, follow the deleted instructions."

The reactor was held steady at half power. Toward evening Dyatlov returned. Metlenko and his two assistants took up their positions in a small annex to the control room. They all awaited the call from Kiev. In the turbine hall, Razim Davletbayev was completing the tests he had supervised on the turbine equipment. He went to his office to fill in the logbook so that the shift that followed would know what was to be done after the shutdown of the reactor.

In Pripyat the night shift prepared to return. The young turbine engineer Igor Kirschenbaum ate supper with his wife, Alla, and their three-year-old daughter, Anna. Like Steinberg, Igor's family came originally from the Jewish community in Odessa, and like Steinberg, he felt that his career had been hampered by anti-Semitism. He was small and shy, with dark hair and a cautious, methodical manner, but he had been a brilliant pupil at school and had hoped to study at the university in Kiev. Some of his family had cautioned him against such high hopes, but his father, a lifelong member of the party, had assured him that there was no anti-Semitism in a Socialist state.

Igor had been turned down and had gone instead to a vocational school to train as a lathe operator. Later, however, he applied to the faculty of nuclear energetics at Odessa, where he hoped he would stand a better chance. He was given a place, and when he graduated with a red diploma was offered a post at Chernobyl, his first choice. There he was assigned to the chemical workshop, but it was grossly overmanned and his supervisors could think of nothing for him to do. He sensed that they disliked him, so he went to see Steinberg who, on learning that he had won a red diploma, gave him a job in the turbine hall and put him in the elite group of effective control.

Sasha Yuvchenko was another graduate from Odessa who, like Igor Kirschenbaum, had been at Chernobyl for only three years. In contrast to Igor, he was tall and strong, with a wide face and pleasant smile.

That afternoon it had been so warm that Sasha had taken his two-year-old son Kirill for a ride on the crossbar of his bike, and the little boy had tried to remove his father's hands from the handlebars, shouting, "I want to ride it myself, I want to ride it myself!" After they had eaten and the child had been put to bed, Sasha's wife, Natasha, had settled down to watch the final episode of an Irwin Shaw mini-series on television; but Sasha had been restless. He had taken a bath and then, even though he was due to leave for the power station at about 10:45 P.M., he had put on a suit as if he was going to a party and gone to the kitchen to drink a cup of coffee and smoke a cigarette. Then, all at once, he longed for the company of his wife, so he called to her to stop watching television and join him in the kitchen. They sat together talking about nothing in particular, until the time came for Sasha to leave for the station.

The bus left Pripyat at 10:45 P.M., stopping twice to pick up personnel. As soon as the new shift arrived at the station, the workers went to the changing room, where each man stored his clothes in a locker, took a shower and, dressed only in his underpants, passed through from the "clean" into the "dirty" zone. Once through, they could not return; there was no handle on the dirty side of the door. Now they put on the white clothes and berets that gave them all the look of surgeons in an operating theater and then proceeded to the final barrier, where they stepped into special boots.

Each then proceeded to his particular post. Sasha went to his office, situated between the third and fourth units, to be briefed by the young man he was replacing, Victor Proskuriakov. He was surprised to discover that the reactor was still running and that the tests on the turbines had been postponed. Victor showed him the program, pointing out that it mostly concerned the turbine engineers. Then, while Sasha went to the third unit to report to his superior, Valeri Perevozchenko, a former officer in the marines, Victor went to the control room to ask for permission to watch the test.

4

There the shutdown was under way. At eleven that night, the load dispatcher in Kiev had given his permission for turbine No. 8 to be detached from the grid. This had been relayed from the head of shift for units 3 and 4, and at 11:10, upon Dyatlov's orders, the shift foreman for unit 4, Yuri Tregub, had started to reduce the power of the reactor from seventeen hundred thermal megawatts to around seven hundred, at which point the test on the turbogenerator could begin.

At midnight the new shift took over. Alexander Akimov replaced Tregub, who, like Valeri Perevozchenko, decided to stay on and watch the experiment. So did the engineer relieved by Igor Kirschenbaum. Above Akimov, in his office in the third unit, was the overall head of the shift, Boris Rogozhkin; under him, in the control room, were, once again, Leonid Toptunov, the young reactor-control engineer; Igor Kirschenbaum, the turbine-control engineer; and Piotr Stolyarchuk, the unit-control engineer. Also present in the control room to observe the test were Akimov's friend Razim Davletbayev, the deputy head of the turbine section; Piotr Palamarchuk, the huge Ukrainian who directed the Chernobyl start-up enterprise, and young Victor Proskuriakov, who had been joined by another young engineer, Alexander Kurdriatsev, eager to add to his experience by watching what was going on.

In an adjacent room, separated from the control room by a window, with their own equipment ranged before them, were Genaddy Metlenko from Donenergo and his two assistants. They were introduced to Igor Kirschenbaum who, although trained for any eventuality, had not been shown the program for the tests. He took his place at the turbine controls, uncomfortable about what he was required to do. He did not like the idea of switching off the turbine and disconnecting it from the circuit while the reactor was still working; he was afraid of the damage it might do to the turbine. However, he was only half the age of Dyatlov, and so assumed that both he and Akimov knew what they were doing. His duty was simply to do what he was told.

At the main control board, Akimov and the young Leonid Toptunov stood facing a bewildering array of dials and gauges. Normally assisted in the running of the unit by a number of automatic devices, they knew that the reactor was running with the reactor's emergency core-cooling system switched off. Assisting them in running the reactor was the Skala computer, but it scanned many thousands of parameters at a leisurely pace and printed out the information at some distance from where they stood.

The most essential information for the young Leonid Toptunov concerned the number of control rods. Power was reduced by lowering them into the reactor, but he had to balance their absorption of neutrons, which reduced the fission, with the naturally dampening effect of the short-lived iodine and xenon gas. To gain more control, in line with the program and with Dyatlov's approval, Akimov now told Leonid to switch off the system of local automatic control, which sent control rods in and out of particular sections of the core to ensure that power never dropped below seven hundred megawatts. To his dismay, Leonid discovered that the power of the reactor then sunk to only 30 megawatts. The reactor had fallen into what was known as an iodine well, from which it was hazardous to attempt to pull it out.

Everything they had been taught now suggested to Akimov and Leonid that the test should be abandoned and the reactor closed down. It was perhaps possible, if the power could be maintained at this level, to wait another twenty-four hours or so for the xenon gas and iodine to decay. Dyatlov, however, had waited long enough. In a state of high exasperation, cursing the operatives for their incompetence, he ordered them to withdraw the control rods to increase the power. Both Akimov and Leonid hesitated: if they did as Dyatlov ordered, the reactor would be working with a dangerously small number of control rods inserted into the core. They argued with Dyatlov until Dyatlov, apoplectic with rage, threatened to bring back the foreman of the earlier shift, Yuri Tregub.

Faced with the direct orders of a deputy chief engineer who had worked with atomic reactors for more than twenty years, Akimov and

Leonid gave way. A further seven control rods were withdrawn, leaving only eighteen in the reactor. This seemed to do the trick. By 1:00 A.M. the power had risen to two hundred megawatts and remained steady. This was below the recommended power level and had been achieved with the same small number of control rods left in the core, but it was enough to proceed with the tests on the turbine.

Soon after 1:00 A.M. the fourth main cooling pump was connected to the heat-transport system. This added to the negative reactivity and meant that Leonid had to withdraw more control rods. Moreover, the low level of power meant reduced resistance to the water flowing through the core; the rate of flow increased, putting pressure on the pumps and causing vibrations in the pipes, and less water turned to steam, which led to a fall in steam pressure in the separator drums.

The operators saw from their dials what was happening and tried to adjust the parameters but did not entirely succeed. Indeed, the steam pressure in the drum separators was falling to a level where it would automatically trigger a shutdown. They therefore overrode the "trip" signals in respect to these variables while Dyatlov telephoned Valeri Perevozchenko, the foreman in charge of the reactor, and summoned him to the control room.

At 1:22 A.M., Leonid saw from a printout of the Skala computer that the reactivity reserve margin was half of what was recommended; in normal circumstances this would have led to an immediate shutdown. He reported the parameters to Akimov at just the moment when the experiment was to begin. At 1:23 the emergency regulating valves to the turbogenerator were turned off so that if necessary the test could be repeated. Metlenko switched on the oscillograph and Kirschenbaum shut off the steam from the turbine. As he did so, Leonid noticed that the reactor's power had begun to rise. He alerted Akimov, who, glancing at the computer printout, shouted to Dyatlov that he was going to shut down the reactor and pushed the emergency AZ button to lower all the control rods into the the core.

There was a thud, followed by further thuds from deep inside the building. At the turbine controls, Igor Kirschenbaum thought the sounds came from the huge tanks of nitrogen installed above the reactor at the same level as the control room. At this moment, Valeri Perevozchenko burst into the control room in a state of great alarm, shouting that while walking along the catwalk high above the top of the reactor, he had seen the heavy caps to the fuel channels jumping up and down in their sockets.

Akimov looked up at the instruments and saw that the descending control rods had stopped. He immediately disconnected the servomotors to let them fall under their own weight. They did not move. Simultaneously there was a terrible tremor, together with a sound like a clap of thunder. The walls shook, the lights went out and a drizzle of plaster dust rained down from great cracks in the ceiling.

At the control panel, the operators looked frantically at their dials and gauges, which were now dimly lit by the emergency circuits. Dyatlov, who had been standing by the electric panel, ran to the main board. He noticed at once that both Akimov and Leonid were in a state of confusion. He looked up at the reactivity gauge; it was positive when it should have been either negative or at zero. Numb with horror, he checked the position of the control rods and saw that they had jammed. He turned and, spotting the two young men who had stayed on to watch the test, Victor and Alexander, ordered them to run to the central hall and lower the control rods by hand. But as soon as they had left Dyatlov realized that it was a nonsensical order because they had jammed still attached to the servomotors. He rushed to the door to stop them but all he could see in the corridor were clouds of dust and smoke.

Dyatlov went back to the control panel, where Akimov had switched on all the emergency pumps to flood the core with water. None of them worked. Both men realized that whatever had caused the explosion—whether it was a separator drum or one of the nitrogen tanks—there was now a real danger of a meltdown. It was vital to get water into the reactor. Akimov turned to Piotr Palamarchuk and Razim Davletbayev and asked them to go to the turbine hall to see what was wrong with the pumps, while Dyatlov ordered Yuri Tregub and Valeri Perevozchenko to open the valves by hand.

In the central hall, before the explosion, Sasha Yuvchenko, accompanied by his immediate superior, Valeri Perevozchenko, had been asked by the operator of the circulation pumps, Ogulov, for some colored paint to mark the different parts of his equipment. The three then went to Sasha's storeroom. There the telephone rang; it was Dyatlov calling

Valeri to the control room. Valeri told Sasha and Ogulov to wait for him in the storeroom.

It was here that they heard a crash that made Sasha think that a crane had fallen onto the building, and a few seconds later they felt a blast of such force that it threw them both onto the floor and blew down the door of the storeroom. When Sasha got to his feet he saw to his astonishment that the concrete walls of his room, a meter thick, had buckled under the force of the explosion.

He staggered out of the room, followed by Ogulov, groping in the dim light and choking in the dust. Ogulov wanted to go back to unit 3, but Sasha warned him that he might be crushed by slabs of concrete falling from the ceiling. "What can have happened?" Ogulov asked him.

"I don't know," Sasha replied. "It may be a war."

While Ogulov went back to his post by the pumps, Sasha tried to ring the control room from his office. He could not get through. Then he got a call from the head of shift in unit 3 asking for a stretcher from the first-aid post in his office.

Confused, because he thought the explosion had come from the fourth unit, Sasha snatched up the stretcher and set off toward unit 3. Before he got very far, he nearly bumped into a man whose face was covered with blisters and blood and whose clothes were filthy and wet. Recognizing him by his voice as one of the pump operators, he offered to help him, but the wounded mechanic brushed him aside, telling him to go to the aid of his companion, Gena, who was still by one of the pumps.

Leaving the stretcher, Sasha ran to the room from which the wounded man had come. It was dark, and at first there seemed to be no one there. Then he saw Gena, filthy and shivering with pain and shock. "Khodemchuk," the wretched man muttered, "help Khodemchuk. He's still trapped up there." Sasha looked up and saw that there was nothing above him but the stars in the sky.

Since Gena could walk, Sasha led him out of the ruins of the pump operators' room and almost at once came face-to-face with Yuri Tregub. "I've been told by Dyatlov to open the valves for the emergency cooling system," Tregub said breathlessly.

Realizing that it required two men to turn the huge valves, Sasha left Gena to go on alone and ran after Tregub. There were two ways of getting to the valves: the nearest one was from above, but there they found that the door had jammed. They went down to the lower entrance; here too the door was stuck, but there was room enough to squeeze through. Water was pouring down from the ceiling and was already up to their knees. They spent some minutes trying to turn the valves, to no avail.

Deciding that they might be able to make out what had happened from outside the building, they exited through a transportation passage and found themselves on the road outside the power station. To their horror, they saw that half of the fourth unit had disappeared. The machine room no longer existed; in its place there was a huge hole, like a burst belly with a mass of torn metal intestines steaming and quivering in the night air. Sasha searched among the debris for the separator drums, but they no longer seemed to be there. Nor could he see the room where Khodemchuk had been stationed. He could make out the tanks that had contained water and nitrogen, hanging like huge matchboxes above the tangle of concrete and steel, and from amid this chaos there came a strange white glow.

Sasha stood still, hypnotized by the extraordinary sight. Then Tregub brought him to his senses. "We're not sightseeing," he said. "We must go back."

Their objective now was to find out what had happened to Khodemchuk, but back in the dim, dust-filled passages of the power station they ran into Sasha's boss, Valeri, with the two young engineers, Victor and Alexander.

"Have you got a flashlight?" Valeri asked.

"Yes, but the battery's low."

"Never mind. Let me have it. We've been told to lower the control rods but we can't get up through the machine room. The passage is blocked."

"The control rods?" asked Sasha. "There are no control rods. We've been outside. The central hall doesn't exist any more."

"An order is an order," said Valeri. "We have to try."

The four of them now climbed from the twelfth to the thirty-fifth level, water from the emergency cooling tanks pouring over them like rain. When they reached the top, where the walls were made of reinforced concrete, they found that the huge door, which itself weighed a ton, had shifted on its hinges. Sasha held it open while Valeri, Victor and Alexander crept through onto the steel rafters, searching for the levers that would release the control rods from the servomotors. To light their way, Valeri took the flashlight from Sasha and handed it to Victor. Then they looked down through the tangle of steam and concrete, and instead of the top of the reactor saw a glowing volcanic crater.

Valeri crept back through the door. "There's nothing to be done," he said.

Victor and Alexander followed. Sasha asked Valeri if he could take a look. Valeri would not let him. "There's nothing to see," he said. He led them back through the smoke-filled passages to the control room to report to Dyatlov.

5

Back in the control room, after the initial panic, a measure of professional calm had returned. Young Igor Kirschenbaum, at the turbine-control panel, had thought first that the roar of the explosion came from the turbine spinning out of control, then that it was an earthquake; he had felt one once in the southern Ukraine. Nevertheless, he remained at his post, switching off all the circuits and waiting for the dials to tell him that the parameters were normal. When this was done, Dyatlov instructed Akimov to dismiss him, along with all other superfluous personnel.

Strictly speaking, Dyatlov was not in charge. Akimov, as head of the shift of the fourth unit was responsible to the overall head of the shift, Rogozhkin, who was in unit 3. Akimov had reported to Rogozhkin over the telephone that there had been a serious accident, but Rogozhkin had not turned up to take charge. Moreover, Rogozhkin knew less about reactors than Dyatlov, and it was Dyatlov who had written the program for the tests on the turbines. It was therefore Dyatlov who took charge, sending Perevozchenko to open the valves and ordering Akimov to start up the emergency cooling pumps. "Lads," he said calmly, "we've got to get water into the reactor."

When Valeri Perevozchenko returned and reported that the reactor had been destroyed, Dyatlov countered calmly that this could not be true. Whatever it was that had exploded, the priority now was to cool the core. He asked Valeri and his three young companions to go up to level 27 and manually open the valves, which would send the water from the turbine hall into the bubbler pool.

Although Dyatlov knew that there had been an explosion, and could see from the dials in front of him that the reactor's parameters were haywire, it did not occur to him that the explosion could have been in the core. He had worked with reactors for twenty years; he had taken numerous courses to bring his initial training up to date; he had studied the voluminous documentation that had accompanied the new RBMK reactors; he had seen the building of the fourth unit and had supervised its commissioning; and never had it been suggested that the reactor itself could explode. He knew that the active zone was dead and realized that the hermetic containment might have been ruptured, but he could not envisage anything worse.

It was time to see for himself. Dyatlov left the control room and immediately ran into three young operators—one, Anatali Kurguz, with skin peeling off all the exposed parts of his body. He ordered him to report at once to the medical unit. He went up to a window that looked out onto the reactor hall and saw that one of the walls had been completely destroyed.

Still convinced that it must have been a burst tank or drum that had caused the damage, Dyatlov ran along the passage to the end of the unit, down the stairs and out onto the road. The first thing to catch his eye were the fires on the roof of the turbine hall. He ran back up the stairs to the control room, which was filled with dust and smoke. He told Akimov to call the fire brigade and turn on the fans to extract the smoke. Then, noticing Leonid Toptunov and Igor Kirschenbaum, he turned angrily to Akimov: "Didn't I tell you to dismiss them?"

"They were dismissed," said Akimov, "but they returned."

Igor explained that he had volunteered to bring respirators from the third block, and that Leonid, feeling responsible as the reactor's operator for what had happened, had come back to help. Dyatlov again told them to leave. Igor obeyed, but once Dyatlov's back was turned, Leonid stayed.

Dyatlov now went back down to the entrance on level 12. Here he could see the damage more clearly. The tanks filled with nitrogen had been destroyed and the emergency core-cooling system would not function. About three hundred square meters of the roof had collapsed, some of the panels and girders crashing onto the two turbines of the fourth unit. At level 6, a pipe had been ruptured, creating a fountain of scalding water, which, falling on torn cables, created a firework display of bright sparks, small fires and minor explosions. Worst of all, from inside the reactor came a terrible glow. It was like Dante's vision of hell.

There were figures in this inferno, operators with fire extinguishers and firemen on the roof. Dyatlov went to the nearest fire engine and advised the young commander where the hydrants were to be found. Then he walked toward the third unit. At his feet there were some smoldering lumps of . . . of what? They looked like graphite, but if they were graphite that could only mean . . . It was dark. How could he tell? Perhaps they were lumps of concrete. He did not pause to inspect them but went back in through a door to the control room of the third unit. There the head of the shift asked him if he should shut down his reactor. Dyatlov said that it was unnecessary, but before returning to the fourth unit he took a dose of iodine from the first-aid kit.

6

Meanwhile, the hefty Piotr Palamarchuk and Razim Davletbayev, asked by Dyatlov to check the pumps, found their way through the dark, dusty passages into the adjacent turbine hall on level 12. The first thing they saw was that part of the roof had collapsed, and even as they stood there pieces of steel and chunks of concrete fell to the ground, together with dripping tar, which formed fires all over the floor. They stayed by the wall, watching the cascade of debris fall on the pipes filled with scalding radioactive water. Then they climbed down to level 5 and stumbled through clouds of steam to look at the pumps. They found the pumps swamped with water; none would function. All around them the ruptured pipes spewed forth their different contents, not just scalding water but hydrogen and oil. The greatest danger came from the hydrogen in the turbines, which could cause a further explosion. Razim began the process of replacing the hydrogen with nitrogen, but first he had to summon electricians to replace the cable that had been destroyed.

Three members of the team that had come from Karkhov to conduct the test on the turbines stood as impotent spectators in the midst of this confusion. Razim told them to leave, and Piotr dismissed three of his own engineers who had been there to measure the vibrations of the turbine during the shutdown. Of the two others, one, Shevchuk, was present in the hall, but another, Shashenok, had been in the reactor hall and did not answer the telephone attached to his post. With Shevchuk, Piotr set off to find him. The corridors were dark; the green lights that indicated a permissable level of radioactivity had turned to red, which made him realize that the explosion had done serious damage, but there was no way of knowing what the levels were and no means of obtaining protective clothing. There was debris everywhere; sometimes they could climb over the fallen girders; at other times their route was blocked, and they had to change direction.

Reaching the reactor hall, Piotr ran into Alexander Kudriatsev, on his way back to the control room, who told him that everything above them had been blown apart. Only now did Piotr realize that the reactor itself might have been damaged, and seeing that he could not reach Sha-shenok's post from where he was, he ran back with Shevchuk to get hold of a dosimetrist, one of the specialists with a Geiger counter responsible for monitoring radiation levels; but when he found one he was told that no one knew the level of radiation because the needles of the dosimeters were permanently fixed at the maximum reading.

Together, the two men set off in search of Shashenok. Once again they had difficulty making their way up the metal staircases and along the concrete passages, at times wading through water that they now realized must be highly radioactive. They reached Shashenok's room. The ceiling had fallen in and scalding water was still pouring out of the pipes, but Shashenok was not there. Then they saw him, lying on his stomach in the passage thirty feet away.

He was barely alive, with atrocious burns over most of his body and several broken bones. Piotr and Shevchuk picked him up, causing him terrible pain, and carried him as best they could along the passage and down the stairs from level 27 to level 9. Shashenok tried not to complain, but he was suffering grievously and asked his saviors to avoid touching certain parts of his body. At times the passage was too narrow for two men to carry him, so the powerfully built Piotr lifted the broken body onto his shoulders.

On level 9 they found a stretcher and used it to carry Shashenok along the passage to the medical station in the first unit. There a doctor gave him a pain-killing injection and sent him by ambulance to Pripyat.

7

On his way to open the valves on level 27 together with Victor and Alexander, the two young engineers, Sasha Yuvchenko remonstrated with his boss, Valeri Perevozchenko. "What is the point, Valeri Ivanovich?" he asked. "I've seen the reactor from the outside. You've looked down into the central hall. It doesn't exist anymore. The right side has been completely destroyed. All the pipes are ruptured; I saw them swinging in the open air. They don't go to the bubbler pool anymore."

"An order is an order," said Valeri again. "Something has to be done."

When they reached the cramped confines on level 27 in which the valves were located, Valeri made the younger men stand aside while he climbed into a dark passage to reach the farthest valves. Alexander followed him. Sasha and Victor started on those closer at hand, but when they turned them a jet of scalding steam blew into their faces; they staggered back and waited for the steam to stop, but it continued to come out of the valve, so they left the room and shut the door.

Valeri reappeared from the passage with Alexander and said, "We've opened the valves, the job is done." As they headed back toward the control room they ran into a dosimetrist in a gauze mask rushing along the passage. Sasha grabbed him to ask him the level of radiation. The man pointed to his meter and said, "It's off the dial," then ran on. With a feeling of dread in his heart, Sasha at last realized that they were all almost certainly doomed to die. His only hope was that he would do his duty courageously in these last few hours of his life.

Now his boss, Valeri, said, "There can be too much of a good thing, Sasha. We've done all we can. Where are the rest of our people?"

"I don't know. I don't know where any of them are."

They came to a pool of light in the passage and Valeri stopped, pulling a damp packet of cigarettes out of his pocket. "Let's rest for a minute," he said, "and have a smoke. Then we'll go and look for the men."

The two men stood silently side by side, smoking their cigarettes. The air was wet with the steam; there were blue flashes from the short circuits. Sasha looked out of the window and thought sadly of his wife, Natasha, and son, Kirill, sleeping in their beds in Pripyat.

IV

1

At fire station No. 2 attached to the Chernobyl nuclear power station, a detachment under Lieutenant Pravik was working a twenty-four-hour shift that was due to end at 8:00 A.M. Leonid Shavrey, a hefty Belorussian from a village fifteen kilometers from Pripyat, was resting on a bunk, waiting to take over as head of the fire-fighting team at 3:00 A.M. His brothers Ivan and Piotr were in the same detachment; they were all delighted to have landed jobs in Chernobyl, thereby escaping the bucolic serfdom of life on a state farm.

At 1:23 A.M. Leonid was awakened by the explosion, which shook the glass in the windows of the fire station. He leaped off the bunk and ran out onto the road. He saw that the roof of the fourth unit was on fire, and that a strange cloud was rising from the reactor. As the alarm started ringing he ran back into the building and jumped up next to Pravik on their new Zil 134 fire engine. As they drove toward the fourth unit, Pravik gave orders through his walkie-talkie to the duty officer at the fire station to alert the whole regional command.

When they reached the fourth unit, Pravik ordered his men to start pumping water from the reservoirs and hydrants onto the building while he and Leonid went to investigate the fire. They tried to go through the passage connecting the third and fourth units to the control room of the fourth unit but were stopped by the fallen panels and broken glass. Then they tried to call the head of the shift on a telephone, but the line was dead. Coming out of the power station, they ran into a distracted operator, who advised them that it would be safer to fight the fire from the other side of the reactor. There they found the second fire engine from Pripyat. While some of the men unraveled the hoses, others went up to fight the fire on the roof of the turbine hall.

At 1:56 A.M. the commander of the Chernobyl fire brigade, Major Teliatnikov, arrived from a house outside the town, where he had been on leave. His initial efforts were to prevent the fire on the roof of the turbine hall from spreading to the third unit. With no special clothing to protect them from radiation, the firemen climbed up the exterior staircases and turned their hoses on the patches of burning bitumen that had been ignited by a shower of hot black rock. The temperature was so high that the roof melted beneath their feet, and as soon as one fire was put out, another would start. After only half an hour or so, the first teams on the roof, led by Pravik and fireman Kibenok, began to feel giddy. Standing on the external staircase seventy-two meters above the ground, Ivan Shavrey felt his legs grow weak and a sweet taste come into his mouth, as if he had eaten some chocolate. He staggered down the stairs. His brother Leonid, too, began to feel weak and sick.

From 4:00 A.M. they were joined by fire-fighting teams from the towns of Chernobyl, Polesskoe and Kiev itself. Finally Leonid Shavrey's unit was relieved and he was told to report to the bunker below the administrative block. As he came in, he recognized Brukhanov, from whom he had once sought a permit to buy a motorbike. Then he heard that his brother Ivan, together with Major Teliatnikov and Lieutenant Pravik, were already in the hospital. When Leonid was checked by a dosimetrist the needle went off the dial. He took a shower and went to lie down in a dormitory with one hundred beds. He had a headache and felt sick but found it impossible to sleep. One of his companions fainted and was taken to the hospital. Leonid asked the doctor what would happen to him and was told that he would be given injections and put on a drip. Terrified at the thought of such torments, Leonid told the doctor that he felt fine.

2

As soon as Rogozhkin was told of the explosion, an alarm call had gone out from his office in unit 3 to a chosen group of people. It was a single uninterrupted ring on the telephone that, when answered, gave a recorded message that they were to report to the power station, where a grave accident had occurred. Brukhanov did not hear it because he was not in his flat in Pripyat. He had had a hard day in Kiev, arguing with underlings of the first secretary of the regional Party Committee about the hay barns he was supposed to have built and about a proposed administrative reorganization that would have removed the town of Pripyat from his control. Only a month before, he had returned in glory as a delegate to the Twenty-seventh Congress of the Communist Party of the Soviet Union with the promise of a decoration as a "Hero of Socialist Labor" when units 5 and 6 were completed; now he was back with the nightmare of serving two masters, the minister of energy in Moscow and the party in Kiev.

Why had he not built the barns? Because he had only managed to get hold of the frames; he was not, after all, a builder but the director of a power station. The argument had gone back and forth all day. When he finally escaped from the offices of the regional committee, he had picked up his daughter and son-in-law in his white Volga and driven with them back to Pripyat. There they had collected his wife, Valentina, and driven on to their dacha for the weekend. Before going to bed, Brukhanov had telephoned the power station to check that the shutdown of the fourth unit was going as planned. He was told that there were no problems and had gone thankfully to bed.

Shortly before 2:00 A.M. he was awakened by a call from Yuri Semyonov, the head of the chemical workshop. "Has no one rung you, Victor Petrovich?"

"No. Why should they?"

"Some sort of accident, something really bad, has happened to the fourth unit."

Brukhanov hung up and tried unsuccessfully to reach Rogozhkin at the power station. Then he dressed, drove back to Pripyat, left his car by his flat and boarded the shuttle bus to the station. As it made for the administrative block, it passed the fourth unit and Brukhanov could see at once that the roof had been blown off the central hall. When he reached his office, he told the telephone operator to summon the entire senior management to the power station. Then he went to the control room of the fourth unit, where he found both Akimov and Dyatlov. He asked them what had happened, but neither could give him a satisfactory answer. Both seemed at a loss but assured him that the reactor was intact.

Brukhanov returned to his office and asked the chief dosimetrist to report on the level of radiation. He was told that their gauges only measured up to 1,000 microrems per second, or 3.6 rems an hour, and in most areas the needles went off the dial. Now Brukhanov telephoned his superiors in Moscow and Kiev. According to the standing orders, he was duty bound to report any accident to a series of high officials, including Gennadi Veretennikov, the head of the All-Union Industrial Department for Nuclear Energy; Grigori Revenko, the first party secretary of the Kiev region; and the Ukrainian minister of power and electrification, Vitali Sklerov. Lastly he telephoned the chairman of the Pripyat Party Committee, Vladimir Voloshko, whom he advised to prepare for an evacuation of Pripyat.

"You must be joking," Voloshko replied.

"I assure you I am not joking," said Brukhanov.

"But an evacuation will cause panic . . ."

Brukhanov hesitated. In theory he had the authority to order the evacuation of Pripyat; in practice, he had never done anything without the approval of the party. He was not a physicist, but he had seen the damage to the fourth unit and heard the first reports of vomiting and faintness among the firemen and the personnel. Yet he had only been officially notified by the chief dosimetrist of the acceptable levels of radiation shown on the dials of the dosimeters, and the decision to evacuate the fifty thousand inhabitants of Pripyat had grave implications. It might cause panic, as Voloshko predicted; it would also break the standing orders that all nuclear accidents were state secrets. It would advertise the accident to the whole world. Who was he, a mere turbine engineer, to make a decision of such a momentous kind?

It was in this frame of mind that he went down to the bunker to meet the assembled leaders of the civil defense.

. . .

Prepared for war, the director of every enterprise from a ministry to a primary school had clearly defined responsibilities in an emergency of this kind. This time the system had gotten off to a bad start. When the duty officer of the civil defense in Kiev had been awakened by telephone and told of the accident, he had asked for the code word, and the operator did not know it. Therefore he had assumed it was a joke and had gone back to sleep. However, the local network had functioned well, and when Brukhanov came down to the bunker he found the heads of workshops, fire brigades, institutes and schools already assembled. There he announced that an explosion had taken place. No one as yet knew what had blown up; possibly it was one of the nitrogen tanks of the emergency core-cooling system. Happily, he said, the reactor remained intact, and every effort was now being made to pump water into it. The level of radiation reported by the chief dosimetrist was at most 3.6 rems an hour.

The head of the station's civil defense, S. S. Vorobyov, objected. He had gotten hold of a dosimeter capable of measuring up to 250 rems per hour and now reported to Brukhanov that in several places, such as the turbine hall, it registered over 200 rems per hour. Surely, with both firemen and operators showing symptoms of acute radiation sickness, the levels of radiation must be much higher than 3.6. Measures should be taken to protect personnel; the other units should be closed down; perhaps Pripyat should be evacuated.

Brukhanov suggested that Vorobyov's instrument might be faulty. He was assured that the reactor was intact, and things were bad enough without creating panic through exaggeration. Stick to the facts as reported by the chief dosimetrist. Above all, no panic.

3

In the control room of the fourth unit, Dyatlov saw that Akimov was ailing and ordered him to send for a replacement. Shortly before five, Akimov telephoned Vladimir Babichev. Babichev called for a car but found a bus waiting at the bus stop. The driver refused to go to the station; he said that that there were roadblocks and that the guards would not let him through. Then the bald head of Nikolai Fomin appeared at the door of the bus; faced with a direct command from the chief engineer, the driver set off for the station.

They were stopped at a roadblock a kilometer before they got there. The guards gave them iodine tablets and were about to send them back, but when they saw Fomin's pass they let them through. Although it was still dark, Babichev could see some of the damage that had been done to the fourth unit. It seemed obvious that something must have happened to the reactor, but Akimov had always told him that it was entirely safe. He went to the changing rooms, passed through the showers, and changed into his white overalls. However, guards would not let him go to the fourth unit but directed him toward the bunker. There he found a crowd of agitated operators and engineers, among them Dyatlov, who seemed dejected and, while smoking incessantly, never met his eyes.

"Ah, Volodya," he said. "Please go and replace Akimov."

"What happened?"

Dyatlov did not answer the question. "Just see if you can get the pumps working."

Seeing a guard in front of the entrance to the fourth unit, and fearing that once again he might not be allowed in, Babichev went to the control room of the third unit and asked the head of the shift, Bagdasarov, what had happened. Again he got an evasive reply; Babichev could not make out whether Bagdasarov was afraid of making a fool of himself, scared of facing the truth, or simply did not know. Babichev went back toward the fourth unit and ran into the deputy head dosimetrist, Krasnozhon. Again he asked what had happened. "For God's sake, put on protective boots and a mask," Krasnozhon said to him. "There's more than a rem a second."

"Was anyone injured?"

"Yes, Sashenok. And we can't find Khodemchuk."

Finally Babichev reached Akimov in the control room of the fourth unit. "What happened?" he asked.

Akimov shrugged. "During the test there was an explosion. We don't know what went wrong."

"I seem to remember you saying that the chances of an accident were one in ten million."

"Yes," answered Akimov, "and this seems to be it."

4

Although Babichev had now replaced him, Akimov did not leave the fourth unit but went with Leonid Toptunov to open the valves that would send water to the reactor. Babichev remained in the control room, asking the operators to report to him, but there was now no clearly established chain of command. In the bunker, Fomin was nominally in charge, but he became increasingly erratic under the strain. At times he was the heroic leader, urging his men to greater efforts; then suddenly his feverish energy would desert him and he would sink into a fit of silent gloom.

To assist Dyatlov there now arrived some of his old friends from Komsomolsk. Vladimir Chugunov, the head of the reactor workshop of the first and second units, was reading in bed when he heard the recorded voice report a serious accident. He telephoned his office and was told that it was in the fourth unit, and that a team from their workshop under Alexander Nekhaev had already left to help.

Anatoli Sitnikov just missed the car that had been sent to fetch Chugunov and had to walk the two kilometers through the forest to the power station. At the station, as the head of a workshop, he was asked to identify the pipes and valves that could be used to pump water to the reactor. He directed the group of young engineers from Chugunov's workshop—Nekhaev, Orlov and Uskov—to the different valves, but the success they had in releasing the water created problems of a different kind. First, the water did not seem to be getting into the reactor, but simply flooded back into the passages and conduits at the base of the station; second, diverting the reserves of chemically clean water to the fourth reactor was leaving the third unit perilously short.

The head of the shift in the third unit, Bagdasarov, asked Fomin if he could close it down, but was turned down. Fomin's implacable bosses in the Ministry of Energy and Electrification had already demanded a schedule for the repair and recommissioning of the fourth unit, so it was unthinkable to tell his superiors that a further one thousand megawatts would now be removed from the grid.

Bagdasarov ignored the chief engineer, and on his own initiative began the process of shutting down his unit, using water from the suppression pool to cool the reactor. By dawn it had closed down.

. . .

At 7:00 A.M. on 26 April, the scientific deputy chief engineer, Mikhail Lyutov, arrived at the station, and at 8:00 A.M., when the new shift arrived on schedule, Victor Smagin came to replace Akimov as the head of the shift of the fourth unit, not knowing that Babichev had arrived earlier. He too had difficulty in getting through the cordon of guards but was reassured by the dosimetrist, Krasnozhon, that the radiation level of 3.6 rems per hour meant that he could work safely for five hours. With the smell of ozone in his nostrils, he went to the control room of the fourth unit, where he found Babichev and Lyutov. Babichev told him that Akimov was off with Nekhaev, Uskov and Orlov working on supplying water to the reactor. Lyutov said that if only he knew the temperature of the graphite in the reactor he would be able to work out what was going on. Smagin told him that from what he had seen on the ground as he arrived, most of the graphite was no longer in the reactor.

Lyutov answered that this was impossible. Smagin led him out into a passage and, with the air burning his throat and his eyes, pointed to the pile of black rock strewn around the ruins of the central hall. Lyutov seemed unwilling to believe that it could be graphite. Exasperated, Smagin pointed out the holes bored in the graphite for the fuel channels. Still Lyutov seemed unconvinced, but he returned to report to Fomin that it was within the realm of possibility that the reactor had been destroyed.

Fomin was equally incredulous. He turned to Sitnikov and asked him, as a nuclear specialist, to find out exactly what had happened to the reactor.

Sitnikov, Dyatlov's oldest friend from Komsomolsk, set out on a tour of inspection. He had arrived at the station soon after Chugunov, and in the ruins of the central hall he had seen Valeri Perevozchenko and Sasha Yuvchenko, who had reported what they had seen. Still, Sitnikov, as an older and more experienced specialist, felt that he must follow Fomin's instructions and take a look for himself.

Deciding that the best view of he could get of the reactor was from the water-treatment plant, he climbed out onto the roof and looked down on the ruins of the reactor. A hot, gaseous blast hit him as he made a mental inventory of the destruction. Much of the piping and many of the tanks that had surrounded the reactor had been blown away by the explosion, and the huge concrete lid—the upper biological shield—had been pushed aside, leaving a crescent-shaped opening to the reactor core, through which came the dreadful glow of a fire.

To a man whose life had been spent cherishing nuclear reactors, this was clear evidence of an unimaginable catastrophe; already choking from the effects of the radionuclides that he had breathed into his lungs, Sitnikov went back to report to Fomin what he had seen. On the way down, he ran into Smagin and told him that the reactor had been destroyed. Smagin noticed that Sitnikov was retching, and that his skin had turned brown.

Down below, knee-deep in radioactive water, the head of the electrical workshop, Alexander Lelechenko, worked with some of his men to repair the electrical circuits and give power to the pumps. Noticing that the pipe bringing hydrogen to the turbines had been ruptured, he had seen the danger of a new fire breaking out in the electrolysis plant. With the help of another electrician, Lopatuk, he had groped his way along the dark, dusty passages to the entrance opposite the hydrogen tanks; there, unsheltered from the burning reactor, they rushed out to close the valves.

Another of the engineers from Komsomolsk, Vadim Grishenka, learned about the accident shortly before he left for work that Saturday morning. He was a handsome man, with a lean, youthful face, as taciturn as his wife, Ylena, was chatty. Their closest friends were the Chugunovs, and it was Nina Chugunov who had telephoned to say that her husband, Vladimir, had been called to the station because of an accident.

On his way to the bus stop, Vadim noticed a tanker sluicing the streets. He thought little of it and went as usual to the fifth unit, where he was deputy chief engineer. The unit was now close to completion, and to try and keep to the schedule construction went on around the clock. As the bus passed the fourth block, he saw the destruction and his heart sank, but it did not occur to him to worry about radiation or to take iodine tablets. However, when he reached his office he called Ylena and told her to shut all the windows and stay in the flat.

As they clocked off, some of Vadim's colleagues from the earlier shift said that they had seen the explosion but had not stopped work. Nor did Vadim or the 268 men on the construction site. At about 10:00 A.M. he was told to report to Brukhanov in the bunker beneath the administration block, and was asked by him to take charge of the third and fourth units since Dyatlov, Chugunov and Sitnikov had all been taken to the hospital in Pripyat. Vadim could see that both Brukhanov and Fomin were frightened and were clearly at a loss about what to do.

In the bunker, throughout that Saturday morning, as more vomiting men were brought to the dormitory, and firemen and operators were being ferried to the hospital in Pripyat, both Fomin and Brukhanov had assured their superiors in Kiev and Moscow that everything was under control. Was it morning or night? Brukhanov had lost all sense of time, but like one of Pavlov's dogs simply reacted to each crisis according to his training. Admittedly there had been an accident, but it could not be serious, because Soviet reactors were totally safe. What had happened? No one seemed to know. What was certain in such circumstances was that secrecy was paramount and that superiors must be reassured.

Brukhanov was not alone in making such judgments; at his side were party officials. The telephone never stopped ringing; he took calls from ministries, institutes, state committees and party leaders. Senior officials appeared in the bunker, among them the head of the KGB in Pripyat and the second secretary of the Party Committee in Kiev. Brukhanov was asked for a written report on the levels of radiation. His advisers prepared a draft incorporating the figures they had been given by the chief dosimetrist, 3.6 rems per hour in the plant and between eight and fifteen millirems per hour in the town of Pripyat.

Brukhanov signed it. It was among his last acts as the man in charge. All around him operators and engineers were doing what they could with no overall sense of direction. With every hour that passed, Fomin seemed closer to disintegration. No one consulted Dyatlov, who ran hither and thither, cringing like a singed cat from the reactor that had let him down. The whole security apparatus had taken on a life of its own, encircling both the power station and the fourth unit, and at times preventing the engineers from getting in.

So, too, had the civil defense; at 5:00 A.M. a general in the militia, Gennadi Berdov, had arrived from Kiev in full uniform to take command. However, neither he, the KGB or the local party leaders dared any more than Brukhanov to make major decisions. None felt authorized to close down the first two reactors, to order the evacuation of the surrounding area or even to warn the people of Pripyat, as they awoke that morning, to remain in their flats with their windows closed. The system of democratic centralism had taught them all that only the all-knowing and all-powerful Central Committee could make decisions of this kind.

Early Saturday morning Brukhanov was informed by Moscow that experts were on their way from the Kurchatov Institute, and that a government commission had been appointed and would arrive in Chernobyl that day. It was to be headed by a deputy prime minister, Boris Scherbina. No decision on evacuation was to be made until he arrived. In the meantime, avoid panic and cool the reactor!

5

Among the recipients of the emergency call at 1:30 A.M. was Vitali Leonenko, the director of the hospital in Pripyat. He was a man with twenty years of experience in health care in the nuclear industry, and he went at once to the power station. As soon as he saw the seriousness of the accident, he telephoned the Third Division of the Ministry of Health in Moscow.

In Pripyat, a general alert went out to all medical personnel. A young surgeon, Anatoli Ben, reached the hospital by 1:50 A.M. He went up to his department on the second floor, changed into his overalls, and then went to the window, from which there was a clear view of the power station a kilometer away. He could see that the roof and the walls of the fourth unit had been destroyed, and that the glow of a fire came from the ruins, together with a thin wisp of smoke. He immediately telephoned his wife, Tatiana, also a doctor, and told her to prepare herself and their neighbors for an evacuation.

Anatoli came down into the reception area just as the first two patients arrived from the station. Both had terrible burns and blisters all over their bodies. One, Anatoli Kurguz, had pushed through powerful jets of scalding steam to close a fire door to protect his friends. He was in a state of shock. He had once been a submariner, and as Anatoli Ben cut off his clothes and bandaged his wounds, he sang songs from his navy days. Every now and then he would stop abruptly and say, "There was a bang, and everything went . . ." Then he would begin to sing again.

Further patients followed in quick succession: the firemen, among them Lieutenants Pravik and Kibenok, and later their commander, Major Teliatnikov. Some were semiconscious; many were vomiting. They were all undressed, washed down, given hospital pajamas and attached to drips. A siren was heard as an ambulance approached at speed. Dr. Ben, who had removed his rubber gloves after bandaging Kurguz, now saw the driver of the ambulance, Gumarov, bring in a severely wounded man. This was Sashenok, who had been carried out of the power station on the shoulders of Piotr Palamarchuk. Ben helped undress him; his body was a raw mess of blisters and burns. Moreover, his rib cage had caved in and his back was so twisted that it was almost certainly broken.

Sashenok was able to talk. "Khodemchuk," he kept saying. "Khodemchuk is still there." He looked in anguish at his wife, a nurse at the hospital, who now stood at his side, but he was so severely injured that he was sent into the intensive-care unit, and there she could not follow him.

Soon after Shashenok came the man who had saved him, Piotr Palamarchuk. He was vomiting and felt faint; he was washed, laid down and attached to a drip. Then came Sasha Yuvchenko, who, though occasionally sick and dizzy, felt unnaturally excited. After he was washed and given hospital clothes, he fell asleep despite his excitement. He was awakened an hour or two later by a nurse in order to be attached to a drip. It turned out that she lived in the same block as he did, so he asked her to reassure his wife when she went home and tell her that he would be home soon.

By about 5:00 A.M. the ambulances were working a shuttle service to bring patients to the hospital. One of the patients was Gumarov, an ambulance driver, who appeared to have been affected by the radiation emanating from the casualties themselves. This made made Anatoli Ben realize just how many dangerous radionuclides must have escaped from the reactor. Some of the patients were well enough to be sent home.

Then Anatoli saw a friend, the head of the electrical workshop, Alexander Lelechenko, waiting for treatment in the reception area. "How are you feeling?" he asked him.

"Not too bad."

"Are they looking after you?"

"Yes. I'm waiting to be sent to one of the wards."

When Anatoli Dyatlov arrived, he refused to be put on a drip and said that he only wanted to sleep. However, the nurse insisted, and after he was on the drip he started to feel better. Indeed, it was common to many of the patients that after the initial dizziness and nausea, they felt reasonably well. It began to look as if only those who had been crushed by falling debris or badly scalded by escaping steam were in any danger. Khodemchuk had been lost in the wreckage of the fourth reactor—no one had been able to reach his body—and at five that morning, Shashenok died.

Inze Davletbayev was telephoned by her husband, Razim, at 6:00 A.M. When he told her that there had been an accident and that he was in the hospital, her legs went weak beneath her. She stood still for a couple of minutes, and then got dressed and went to the hospital. There she found that no visitors were allowed. She asked after her husband, but was told that he was not there. She wandered around to the back and stood looking up at the window: suddenly, she saw Razim and he saw her. He waved, and the others in the ward waved too. They all seemed cheerful, and so Inze felt enormously relieved. She stood looking up at Razim's curly black hair, thinking how handsome he looked in the weak sun of the dawn. Beside him was a doctor with whom they had often played tennis, and from the window Razim shouted down that he would send out a letter. She went back to the entrance; the note from the doctor warned her to return home, remove her shoes, close all the windows and keep the children indoors.

Valentina Dyatlov had been woken by the emergency call at 2:00 A.M. Since her husband was already at the station, she hung up the telephone and went to sleep. At a quarter to seven, Dyatlov himself telephoned from the hospital. He said she should not worry, that he was well enough, that he was going to rest for a while and would see her soon.

At 7:00 A.M. Valentina set off with her granddaughter to walk to their dacha outside Pripyat. It took them forty minutes to get there. They stayed there working in the garden, planting onions and peas. Then her daughter, the little girl's mother, appeared. "Apparently there's been a serious accident at the station."

"Nonsense. Look. It's perfectly all right." They could see the huge white power station from the dacha; only the fourth unit was shielded from their view. "Anyway," Valentina added, "I've spoken to Anatoli. If anything serious had happened, he would have told me."

Later that morning, Valentina went to the hospital in Pripyat with some cigarettes for her husband. She was not allowed in. She waylaid an oculist she knew, asking him to deliver the cigarettes, but he said he would not be admitted to Dyatlov's ward. Finally a nurse agreed to smuggle them in.

Going round to the back of the hospital, Valentina joined the group of women waving to their husbands at the window. She saw Dyatlov, who shouted down to her to send in some milk, cucumbers and mineral water. She rushed back to her flat to get them, and then returned to the hospital where a woman with a granddaughter at the same kindergarten as Valentina's smuggled in a rope which Dyatlov lowered from the window of the ward to lift up the bag with his supplies.

Nina Chugunov knew of a better antidote to radiation than milk, cucumbers or mineral water. Since the early days in Komsomolsk, it had been believed that alcohol washed radionuclides out of the system. As luck would have it, stocks were high; she had laid in a good supply for the May Day holiday. Moreover, the Komsomolsk group were great friends of the Bens; they had all celebrated International Women's Day together the month before, when Dyatlov had recited poetry and they had all had a cheerful time. Therefore Ben was unlikely to object to the bottles that were later raised in the basket; indeed, the hospital itself had supplies of pure alcohol.

In the ward, Chugunov shared the vodka that his wife had sent in. Dyatlov came over, then Sitnikov and Akimov, and as they drank the supposedly cleansing alcohol they discussed the accident and tried to work out what could possibly have gone wrong.

Among the firemen, too, alcohol was supplied for medicinal purposes by a nurse who happened to be Ivan Shavrey's wife. Ivan shared it with a friend; they each drank fifty grams. At first it seemed to make them better, but then they became sick. As soon as he heard of the accident, Shavrey's father came from the state farm on the other side of the Belorussian border with supplies of fresh milk, but that too made them vomit, and a bitter folk medicine, smuggled in by Ivan's wife, did no better.

At 10:00 A.M. on 26 April, while the victims of the accident administered these homespun remedies, the team of experts arrived from Moscow. It was headed by Georgi Seredovkin, a specialist in radiation illness. He knew only too well that appearances could be deceptive, and began collecting blood samples from the 120 patients so that through biological dosimetry he could judge which of them should be sent for special treatment in Moscow.

6

Despite the shuttle service of the ambulances and the increasing activity around the hospital, few awoke that morning in Pripyat with any premonition other than that it was going to be an exceptionally lovely day. The sun shone as strongly as it usually did in June, and the prospect of a weekend followed by the May Day holiday put the people of Pripyat in a festive mood. As was usual on a Saturday morning, the children set off for school while their mothers went shopping. Those who could left early for their dachas in the country.

Lubov Kovalevskaya, the journalist whose article in *Literaturnaya Ukraina* had so outraged the management of the power station, had stayed up late the previous night working on a poem about Paganini. While writing, she had played his music, and at around midnight her work was completed. Exhausted, but with the phrases still spinning around in her head, she had taken a sleeping pill and gone to bed.

When Lubov awoke the next morning her mother told her that there had been two explosions during the night from the direction of the power station. Wondering whether this had been more than a simple release of surplus steam, Lubov got dressed and walked toward the power station. She got as far as the town hall when she suddenly suspected that something more serious might have happened; there was a whiff of ozone in the air and a taste of dust in her mouth. A group had gathered in the square in front of the offices of the city soviet. They asked what had happened and were told that there had been a fire at the station. Lubov went to her office and telephoned the Party Committee to ask for confirmation.

"Who has been saying this?" she was asked. "Do you want to cause panic spreading rumors of this kind?"

Now Lubov went with another poet to the edge of the town, where on a raised piece of ground they could get a good view of the station. Paganini's music was still ringing in her ears as she looked with increasing dismay at the ruined walls of the fourth unit. She was terrified by what she saw, but she could not turn away. At the same time she felt as ecstatic as she had when writing her poem. A bright sun shone in a clear blue sky; everywhere people were enjoying the radiant weather—swimming, sunbathing, boating, fishing, sitting in cafés, going on picnics, even holding a wedding in the open air—while she and her fellow poet stood mesmerized by the sight of an unimaginable disaster.

On the way back to her flat Lubov passed policemen with walkie-talkies and tankers sluicing down the streets. When she got home, her daughter had returned from school and Lubov told her to remain indoors.

Another Lubov, the wife of Alexander Lelechenko, had been awakened at 2:00 A.M. by the wife of a friend. "Haven't you heard what's happened?" she asked.

"No."

"Something exploded in the turbine hall. My husband went pale when he heard, and ran off to the station."

Lubov Lelechenko immediately rang her husband's office but was told he was not there. She reached the central switchboard and eventually found him. "What's happened?"

"Nothing. Why?"

"I heard that there's been an accident."

"Don't worry. I'm fine."

Lubov Lelechenko could not get back to sleep and waited for the dawn. At 7:00 A.M. she set off for the primary school where she was the headmistress and teacher of math. Normally, she passed old peasant babushkas selling carrots and radishes, but this morning there seemed to be few of them around. She was the first to arrive at the school building, and because it was stuffy she opened all the windows. Later, a lady janitor appeared and asked for her son. "They're going to evacuate all the children and I wanted to take my son out of school because he isn't well."

"But what happened?"

"The reactor exploded. People have been killed."

The telephone rang. It was the head of the educational service, who told Lubov Lelechenko to close all the windows of the school. "The children must be kept indoors. Wash the floors. Put wet cloths across the entrances to keep the air out."

Soon the children trooped in, prattling rumors. Some said their fathers had been working on the construction site of the new reactors during the night and had seen the explosion. Lubov Lelechenko went to her office and telephoned her flat. Her husband had returned. "They say the reactor has exploded," she said.

"No, no."

"Are you telling the truth?"

There was a pause; then he said, "I'm tired. I'm going to sleep."

Lubov was impatient to get home. She bought some food from the school canteen and hurried back to the flat at 3:00 P.M. It was dark inside; the curtains had been drawn. She went to the kitchen and prepared some lunch, then crept into the bedroom. Her husband was awake and looked worn and troubled. She offered him lunch. He said he was not hungry but would like a drink. She brought him some mineral water and put it next to his bed. A few moments later, when he came into the kitchen, she saw at once that the skin of his face, his hands and his legs were all red, and that the flesh on his hands was swollen. She imagined that he had washed himself at the power station and had scrubbed too hard. She asked him how he felt; he said it was as though he had spent too much time in the sun.

"Shouldn't you see a doctor?"

"I went to the hospital. They took a blood test, but I got fed up with waiting."

Lelechenko went back to bed but could not sleep. At four in the afternoon, he got up to go back to the station.

"Has there been an accident?" she asked.

"Yes. The reactor has exploded."

"Please stay."

"No. I must go back to the station. My men are still there."

• • •

Women also went to the power station for the eight o'clock shift. On her way to her workshop, one looked in at the medical center for some cotton wool because of an unusually heavy flow of menstrual blood. As she came out, she dropped it in the hallway; she picked it up, brushed off the dust, and went to work.

Nina Chugunov was awakened at 6:00 A.M. by a friend who worked in a storeroom at the power station. When she heard that there had been an accident, she began to cry, but an hour later she sent her son off to school. She telephoned her friend Ylena Grishenka whose husband, Vadim, deputy chief engineer for reactor No. 5, had left as usual for work. Like Nina, Ylena was anxious for her husband and sent her daughter, Alyona, off to school.

There all the windows had been opened because of the glorious weather but halfway through the morning were closed. Iodine tablets were given to the pupils, which made some of them vomit. One little girl started to cry when she was told by a fellow pupil that her father had been killed in the station; her friends tried to comfort her and then she was told by her teacher that it was untrue.

Lessons continued as usual, but before leaving for home the teachers told the children to cover their mouths and noses with their hands on their way home. Once out in the street, however, they saw young mothers pushing their babies in prams and younger children playing in sandpits; as Alyona reached her own block she saw her mother on the balcony, hanging the wash out to dry.

Natasha Yuvchenko had slept badly that night. It had been stuffy and Kirill had kept crying; she could not seem to calm him down. She went into the kitchen, then back into the bedroom, quieted Kirill for a moment and snatched some fitful sleep. Finally she got up at 7:00 A.M. and was getting ready to go to work when the nurse who lived on their block came back from the hospital. She told Natasha that there had been an accident at the station and that Sasha was in the hospital. Natasha swayed; she felt weak. The nurse assured her that it was nothing serious, but Natasha wanted to find out for herself. She left Kirill with her neighbors and ran to the hospital. She saw Sasha at the window and he waved to her. She went to the entrance but was not allowed in.

. . .

From the hospital, Dr. Ben looked out with growing alarm at the people of Pripyat going about their business as if nothing had happened. He could see a young man stripped to the waist strumming his guitar in the sun, and across the river people were sunbathing on the beach. He telephoned his wife, Tanya, for a second time to ask her to warn her neighbors of the danger. She went downstairs to give some iodine to the family who lived below and to tell them to keep their children indoors. One of the women laughed. "I know all about the accident. My husband was fishing in one of the reservoirs. He saw it happen, and he's brought a whole load of radioactive dirt back into the flat."

Later that afternoon, Dr. Ben again looked across toward the river to the café by the jetty. There sat a young bride and groom, surrounded by their family and friends, celebrating their wedding. He slipped out of the hospital and crossed to the café to warn them about the radiation. The young people smiled. They looked at the clear blue sky, the gentle sun, the lazy flow of the river. "It could be dangerous," Ben repeated, and ran back to the hospital. When he looked down again from the window of the ward on the second floor, the wedding party was still where he had left them, laughing and raising their glasses to toast the happy couple.

7

Georgi Seredovkin, who with his team of specialists was now measuring the radiation received by the casualties from the power station, had been sent from Moscow's Hospital No 6. Like the hospital in Pripyat, this came under the secret Third Division of the Ministry of Health; however, the department on the top floor, which specialized in radiation-related accidents, was in fact the in-patient department of the Institute of Biophysics, financed and administered by the Ministry of Medium Machine Building.

The director of the institute was Academician Leonid Ilyn, who as a young man had done his research in the aftermath of the terrible accident at Mayak, and the head of the institute's in-patient department at Hospital No. 6 was a redoubtable woman of sixty-two, Angelina Guskova, who thirty years before had been Igor Kurchatov's personal doctor. A convinced Communist as well as a dedicated physician, she had devoted her life to her country and her calling. Unmarried, she was recognized both at home and abroad as one of the world's leading authorities on radiation sickness.

Within an hour of the accident at Chernobyl, Dr. Guskova was summoned by the duty officer to the Ministry of Health. A car was sent to fetch her from Hospital No. 6, where she lived, and half an hour later she was conferring by telephone with the director of the medical services in Pripyat. The information she received was confusing; on the one hand she was told that the accident was not serious—a minor explosion and a fire—but on the other the number of casualties kept rising. The symptoms reported by the doctors suggested that some were victims of radiation sickness, yet the information from the power station was that levels were relatively low.

By 5:00 A.M. it had become clear to her that a considerable number of people were seriously ill. She returned to Hospital No. 6 and asked the director to evacuate patients from the other wards; she might need all of the hospital's six hundred beds. Then she alerted the necessary medical personnel, and Dr. Seredovkin and his team of experts left for the airport at 7:00 A.M.

The alarm that went out on the reserved telephone circuits to the Ministry of Energy and the Ministry of Health was also received at the headquarters in Moscow of the civil defense: the code "one-two-three-four" meaning nuclear-radiation-fire-explosive accident. General Altunin, the commander, was in Lvov, summoned to a conference along with other senior officers of the Soviet armed forces by the minister of defense, Marshal Sokolov. A call was therefore made at 3:00 A.M. to his deputy, General Ivanov. In the absence of General Altunin, he was in charge.

The son of a brewery worker in Leningrad, Ivanov was a jovial, thick-set officer in the Zhukov mold. He had left school at the age of seventeen to enroll in the tank academy and prepare for war. In 1940, as a young lieutenant, he had taken delivery of a tank from the factory in Stalingrad. Four years later, after being wounded on four separate occasions, he fought in the final offensive on Berlin.

Now, as deputy head of the civil defense, he moved into action. He sent for a car to take him to civil defense headquarters, and there summoned their expert on nuclear matters, General Maximov. Of course General Ivanov had his own views on the dangers of radiation; he had served for a while at the nuclear testing grounds in Kazakhstan and had received a fair dose of radiation, but nothing, he thought, that could not be flushed out by a bottle of vodka followed by a good sweat in a sauna.

Ivanov left for a military airport outside Moscow where an A-26 was waiting to take off, but there was told to wait for experts from the Ministry of Energy and the Kurchatov Institute.

An emergency call came through to the Central Command Post at the Ministry of Defense about an hour after the accident occurred. The duty officer immediately contacted the chief of the general staff, Marshal Sergei Akhromeev, to report the explosion, which had possibly released radionuclides into the atmosphere. At 3:30 A.M. Marshal Akhromeev called a conference of senior officers at the Ministry of Defense. There was still no accurate information about what had happened, but Akhromeev deemed the situation serious enough to send a mobile radiation reconnaissance unit to the area, and to direct military transport aircraft to the Privolzhsky military district near Kuibyshev on the Volga. There a specialized detachment of the chemical troops was stationed who were equipped to deal with nuclear accidents—in particular the inadvertent re-entry of satellites powered with nuclear reactors.

At 6:00 A.M. Akhromeev was told by the commander of the Kiev Military District that while the fires had been extinguished on the roof of the reactor, the explosion appeared to have occurred in the reactor itself and that the situation was now out of control. With the accident appearing more serious by the hour, General Akhromeev now ordered a full military alert, and the entire general staff was summoned to the command center. The management of the railway system was ordered to organize rolling stock in Kuibyshev to transport the heavy equipment of the specialist brigade. Helicopters and specially equipped reconnaissance planes were ordered to Chernigov, the nearest airport to Chernobyl. A report was made to the minister of defense in Lvov, and finally Akhromeev ordered the commander of the chemical war division of the Soviet armed forces, Colonel General Pikalov, to go at once to the scene of the disaster.

. . .

Armen Abagyan, the jovial Armenian who directed the All-Union Research Institute for Nuclear Power Plant Operation (VNIIAES), was awakened at 3:00 A.M. by a call from the Ministry of Energy and Electrification. Despite the code, he asked the operator whether this was not in fact a civil defense exercise. She said it was not, so he quickly packed a bag and drove to the ministry. Some minutes later, the minister, Mayorets, arrived together with Veretennikov, the head of the All-Union Industrial Department for Nuclear Energy. Sitting around the minister's desk, they put a call through to Chernobyl. Brukhanov spoke to Veretennikov and reported that there had been an explosion but that they did not yet know what had caused it. The local fire brigade was dealing with the fire in the turbine hall and all possible measures were being taken to cool the reactor.

Despite Brukhanov's reassurance, there was a growing sense at the ministry that an extremely serious accident had occurred. A team was chosen to fly to Chernobyl at once: the minister of energy and electrification, Mayorets; his shadow in the Central Committee, Vladimir Marin; Slavsky's deputy at the Ministry of Medium Machine Building, A. G. Meshkov; Nikolai Poloshkin from Dollezhal's institute, NIKYET, which had designed the reactor; and Armen Abagyan from VNIIAES. Professor Ryzantzev, the head of nuclear safety, was summoned from the Kurchatov Institute, and attempts were made to find Dr. Kalugin, the institute's expert on RBMK reactors. Kalugin could not be found, and Ryzantzev arrived late for the meeting because he had gone first to the Ministry of Medium Machine Building. Shortly before 10:00 A.M. the group finally assembled at the military airport, where an impatient General Ivanov was waiting for them. There were two women in the group, which was unusual when there was danger, but they were needed to take samples of blood. At 10:00 A.M. the A-26 took off for Kiev.

8

Valeri Legasov awoke that morning unaware that the accident had occurred. He had the choice, on a Saturday, of going to his department at the university, staying at home with his wife, Margarita, or attending a meeting of his Communist party cell at the Ministry of Medium Machine Building. True to his ideological outlook, he chose to go to the party meeting. He was wearing a dark suit and, despite the warm weather, took along his smart leather overcoat.

The meeting was held in the office of the octogenarian minister, Efim Slavsky, who told Legasov when he arrived that there had been some sort of accident at the Chernobyl power station. Then they sat down to listen to a report by the party secretary of the Kurchatov Institute, Andrei Gagarinski. Legasov had to stifle a yawn. Reports of this kind were always a catalog of victories and triumphs, though Gagarinski did mention in passing the accident at the Chernobyl station which of course was not the responsibility of their ministry, the Ministry of Medium Machine Building, but that of the Ministry of Energy and Electrification, and in any case was not of sufficient gravity to affect the glorious future of nuclear power.

At midday there was a break, and Legasov went up to the offices on the second floor to find out more about the accident. There he was told that the government had already appointed a commission to take command, that he was a member and that he was to fly to Chernobyl from Vnukovo Airport at 4:00 P.M.

The chief of the nuclear safety department at the Kurchatov Institute was Professor Eugene Ryzantzev, a tall man with a deep voice, but his speciality was the small VVER reactor, used for nuclear icebreakers and submarines. The expert on RBMK reactors was Dr. Alexander Kalugin, a genial man with a pudgy face who, when he awoke that morning and saw what a fine day it was, decided that it was the perfect occasion to let his wife stay in bed and to take his four-year-old daughter for a long walk. They went farther than he had intended, and she toddled very slowly, stopping every now and then to look at a dog or balance on a wall.

When they returned home at about 1:00 P.M., Dr. Kalugin's mother-in-law, who lived with him, complained that the telephone had never stopped ringing and that he was wanted by his office. Kalugin immediately phoned a deputy director who said that a car was waiting for him at the entrance to the block of flats. Twenty-five minutes later he was at the Kurchatov Institute, where he was taken straight up to Legasov's office.

"There's been a terrible accident at Chernobyl," Legasov told him, "the worst the world has ever known. Apparently the fourth reactor has exploded."

Kalugin was dumbfounded.

"I have to be at the airport at four," said Legasov, "but I'd like your opinion on what might have happened."

Kalugin was confused. "Without more information I don't know how I can even make a guess."

After Legasov departed Kalugin went to the Ministry of Energy and Electrification, where in Veretennikov's office he found his boss, Professor Ryzantzev, and another colleague from the Kurchatov Institute, Dr. Konstantin Fedulenko. As scraps of information came in from Chernobyl, they discussed what could possibly have happened. Apparently the operators had been conducting some sort of experiment and many of the safety devices had been disconnected. Still, they did not see how an RBMK reactor could actually have exploded: a VVER, perhaps, but not the workhorse of the industry, the RBMK. It seemed more likely that there had been an explosion of hydrogen in the bubbler pool, although even that seemed unlikely. Could it have been sabotage? Or were the operators so inadequately trained that they . . . What? The worst they could envisage was the kind of mishap that had occurred in 1984—a blockage in the supply of water leading to a meltdown of one or two fuel elements—but not an explosion.

As dusk fell reports came in of numerous casualties arriving in Hospital No. 6. Then at 10:00 P.M. a call came through from Poloshkin at Chernobyl. He had been above the reactor in a helicopter and could confirm that it had been completely destroyed, and that the graphite stacks were on fire. Blocks of graphite and nuclear fuel had spewed out around the power station, millions of curies of radionuclides were rising into the atmosphere, and preparations were under way for the evacuation of Pripyat.

It was immediately decided that a second team would leave for Chernobyl the next day, including Ryzantzev, Kalugin and Fedulenko. They all returned home, but when Fedulenko reached his flat, the telephone rang; it was Alexandrov's secretary at the Academy of Sciences. A moment later Fedulenko heard the old man's tremulous voice. "Were you at the Ministry of Energy? What did you find out?"

"It's still not clear," said Fedulenko, afraid to alarm his revered director, who he knew was grieving over the imminent death of his wife. "We are waiting for further information. But they are drawing up plans to evacuate Pripyat."

"Ah," said the old man in a quavering voice. "Then it must be serious. Yes, it must be very serious indeed."

V

1

At 6:00 A.M. that Saturday, the Soviet prime minister, Nikolai Ryzhkov, was called at home by the minister of energy, Mayorets, and told about the accident at Chernobyl. There had been a fire; no one knew what had caused it; experts from the ministry were on their way; everything was under control.

To the good-looking and relatively youthful prime minister, with his thoroughly decent manner but slightly lugubrious expression, an accident in a nuclear power station was now added to a long list of other preoccupations: the war in Afghanistan, the campaign against alcoholism starting on 1 May, the American "star wars" program and the Herculean tasks imposed by *glasnost* and *perestroika*. Ryzhkov was relatively new to political power—like Gorbachev, he had been brought onto the Central Committee by Andropov only five years before—and the ladder he had climbed was not that of the party apparatus but that of industrial management and economic planning. At one time a miner, later an engineer, he had run the gigantic Uralmash military-industrial conglomerate in the 1970s and was as close to an apolitical technocrat as it was possible to be in the ideologically oriented Soviet system.

When he got to his office at nine—it would have been earlier on a weekday—Ryzhkov heard again from Mayorets. It now looked as if the accident was more serious than at first supposed. The fire was not just in the turbine hall, but had followed an explosion in the reactor itself. Although an experienced engineer, Ryzhkov knew nothing about atomic physics, and so ordered Mayorets to send out experts while he formed a commission of inquiry. The obvious man to head it was Ryzhkov's dynamic deputy, Boris Scherbina, the minister responsible for fuel and power. Learning that he was in Orenburg on the Ural River, Ryzhkov reached him by telephone and told him of his appointment. He was to wind up his business in Orenburg as quickly as possible and go to Chernobyl. At 11:00 A.M. Ryzhkov signed the decree establishing the government commission.

The son of a railway worker from the Donetsk region of the Ukraine, Boris Yevdokimovich Scherbina had risen through the party apparatus in Kharkov to become first secretary of the industrial region of Tyumen, and from 1973 to 1984 was the minister of petroleum and gas industry enterprises. In this capacity he had been responsible for building a 4,500-kilometer pipeline to bring natural gas from western Siberia to be sold for hard currency in the West. Bedeviled by the difficulties endemic to any enterprise of this size, he also had to deal with a U.S. government ban on the sale of any American components for the pumping stations, whether manufactured in the United States or Western Europe; further-more, the story was put out in West Germany that he was using convict labor from the gulags.

A small, wiry man, Scherbina had been rewarded for his achievements with four Orders of Lenin, a Hero of Socialist Labor, an Order of the October Revolution and promotion in 1984 to deputy prime minister, with responsibility for energy throughout the Soviet Union.

By noon that Saturday, having finished his business in Orenburg, Scherbina flew back to Moscow and went home to change his clothes and have lunch. At 4:00 P.M. he was back at Vnukovo airport, and with the other members of the commission, who had been waiting for a couple of hours, took off for Kiev. Knowing little about atomic physics, Scherbina invited Valeri Legasov to sit next to him on the plane, and asked him what might have happened. Legasov was unsure. He told Scherbina about the accident at Three Mile Island, but explained that a similar mishap was unlikely because the reactor at Chernobyl was of a different design.

When they landed at Kiev's Juliana airport the commission was given a reception appropriate to Scherbina's rank: a delegation of local party leaders was waiting on the tarmac in front of a long line of huge black

Volgas and Zils. The Ukrainian Communists all looked anxious: because Chernobyl was an All-Union enterprise, they knew even less about what had happened than Scherbina and Legasov. They sped north from Juliana with a motorcycle escort provided by the militia, through the towns of Dymer, Ivankov and Chernobyl itself to the party building in Pripyat, which they reached at 8:00 P.M.

An operational headquarters had been set up in the office of the town party secretary, Gamanyuk. Two hours before, the minister, Mayorets, had chaired a meeting at which he had not only refused to countenance the shutdown of the first two reactors but had ordered the wretched Brukhanov to draw up a schedule for the recommissioning of the fourth unit by the end of the year so that Mayorets could placate Scherbina.

Scherbina was small, but like Napoleon he made up for his insignificant stature with a tough, determined manner, thin lips fixed in a hard expression and eyes that mesmerized his subordinates like mice before a snake. His manner terrified Brukhanov and offended the valiant General Ivanov, who felt he was being treated like a fool.

Scherbina asked for reports from the assembled experts and officials and listened with menacing silence to Mayorets, the minister of energy; Marin, responsible for atomic power in the secretariat of the Central Committee; Yevgeny Vorobyov, a deputy minister of health; Vasili Kizima, the head of construction at the Chernobyl nuclear power station; Gennady Shasharin, Mayorets' deputy; and Boris Prushinsky, the chief engineer of the State Committee of Atomic Energy, who with Nikolai Poloshkin from NIKYET had flown over the damaged reactor at a height of eight hundred feet in a helicopter belonging to the civil defense.

They told him what they had seen. The explosion had destroyed the reactor and ignited the graphite in its core. The upper biological shield, weighing one thousand tons, had been blown to one side, leaving the inside of the reactor open to the sky. It was red-hot from the graphite fire. The pumps and piping of both the main circuit and the emergency supply had been destroyed; so too were the separator drums and auxiliary tanks. It was clear that if any water was actually getting to the burning graphite, it was having no effect; if anything, it was making matters worse by creating radioactive steam.

Next came a report from the commander of the chemical troops,

Colonel General Pikalov. A cossack from the Kuban, tall and vigorous with strong features, a florid complexion, thick, bushy eyebrows and a deep voice, he talked with the assurance of someone on his home ground.

Pikalov was the son of a cavalry officer and had served in the army since the age of sixteen when his class volunteered as snipers as the Germans approached their school in Rostov-on-Don. After Rostov, he had fought in Moscow, Voronezh, Stalingrad, Minsk, Berlin and Prague. He had been wounded three times (marrying the nurse who had tended him on the last occasion), and still had twenty fragments of shrapnel buried in his head. In 1945, he had joined the newly formed Military Academy of Chemical Warfare. Thirty years later, after a spell on the general staff, he had been put in command of the chemical warfare division of the Soviet Armed Forces. Earlier that Saturday, he had been attending the conference called by the minister of defense in Lvov. He had flown directly from there to Chernobyl in an Mi-8 helicopter, and had at once circled the reactor to assess the damage. Appalled by what he had seen, he had radioed Marshal Akhromeev in Moscow to alert units in Ovruc and to send specialized equipment from Kiev.

Pikalov now gave Boris Scherbina a preliminary assessment. As a result of the explosion in the fourth reactor there had developed two exceedingly dangerous sources of radioactive contamination. The first was the vapor rising from the reactor and forming a cloud that was drifting north-northwest; at midday, they had already measured a level of thirty-nine roentgens an hour fifty kilometers from the station. It was extremely fortunate that this highly radioactive vapor had been carried away from Pripyat by the prevailing winds, but Comrade Scherbina should bear in mind that the wind could change.

The second source of radiation, Pikalov explained, was the debris scattered by the explosion around the reactor. This was largely graphite, but there were also fragments of zirconium from the fuel-assembly casing, and even deposits of the uranium fuel itself. This second source made it extremely hazardous to approach the reactor for anything other than a very short period of time.

Dr. Abagyan confirmed what Pikalov had said. This robust Armenian, director of VNIIAES, the All-Union Research Institute for Nuclear Power Plant Operation, had arrived that afternoon with dosimeters capable of measuring any level of radiation. He described how he and his assistants had at once tried to get through to the fourth unit. They had found the corridors flooded with highly radioactive water and therefore had climbed up to a window in the third unit that looked out over the damaged reactor. It was apparent from what they saw that the explosion had taken place in the core itself, and that a fire still raged in what remained of the graphite stacks. They had looked out for less than a minute, and in that short space of time had measured a dose of fifteen rems.

Scherbina remained calm. From what he had been told by these and other experts, it was apparent that they faced an accident of a catastrophic and unprecedented kind. A number of important decisions had to be made at once. The simplest was made first: the first two units, as well as the third, were to be closed down. The second question involved the evacuation of Pripyat. Scherbina turned to General Berdov, the deputy minister of the interior of the Ukraine and commander of the militia. He had been at Chernobyl since 5:00 A.M. and now reported that contingency plans had been drawn up to evacuate the town. Six hundred Icarus buses had been requisitioned from the city of Kiev, and a further 250 buses were available, as well as 300 cars and two trains. Already a number of families had left in their own cars, whether fleeing from the accident or for the weekend it was hard to tell. Roadblocks were now in place to prevent cars from approaching Pripyat. From the point of view of the civil defense, the greatest danger was mass panic, leading to a disorganized exodus through possibly contaminated territory. For that reason, Berdov had ordered his men not to wear masks or respirators in the town; there were not enough for the whole population, and to distribute some to the few would cause panic among the many.

Next, General Ivanov reported on radiation in the town. Ten different samples showed levels of between one and fifty milliroentgens per hour. However, the question of evacuation was legally the responsibility of the Ministry of Health. Scherbina turned to its representative, who took the view that evacuation was unnecessary. The established norms were as follows: at seventy-five rems, evacuation was mandatory; at twenty-five rems it was unnecessary; between twenty-five and seventy-five, it was a matter of judgment by the local authorities. Clearly, if present levels remained constant, the population was in no danger of receiving even the minimum permitted dose of twenty-five rems. The principal danger came from the short-lived radioactive isotopes of iodine 131, which would lodge in the thyroid gland; this was particularly hazardous for children. Therefore stable iodine should be distributed to saturate their thyroid glands and allow the radioactive iodine to pass through the body.

The physicists disagreed. They argued that it would be best to play safe. No one knew what was going on in the reactor. There could possibly be a power surge or a meltdown, and the direction of the wind could change. To have to evacuate the people of Pripyat at a later date under a cloud of highly radioactive aerosols would subject them to greater risk.

But other factors had to be taken into consideration. Could the evacuation of fifty thousand people be concealed from the rest of the population? It would not be reported by the media, of course, but rumors would quickly reach Kiev. Would these cause panic and a mass exodus of three million people from the city? And how would this look to the outside world? What effect would it have on the prestige and standing of the Soviet Union? Certainly there might be some casualties if they stayed, but they would be trivial compared to those of the Great Patriotic War.

For once the decisive Scherbina did not make up his mind. He would have to get further advice. General Berdov was to deploy the buses for an immediate evacuation; a final decision would be made in the morning. In the meantime the commission was to be divided into different teams to undertake specific tasks. Dr. Abagyan was to collate data on the levels of radiation—to discover, above all, whether there was evidence that the reactor was still active. Academician Legasov was to suggest methods of limiting the effects of the accident. All were to report back when the commission reconvened at 7:00 A.M. the next day.

2

Boris Scherbina now decided to take a look at the reactor for himself. With Valeri Legasov and Mayorets' deputy, Shasharin, he took off in a helicopter from the square in front of the offices of the Central Committee and flew toward the red glow that lit up the night sky.

They were horrified by what they saw. In contrast to the clean, white exterior of an undamaged nuclear facility, the fourth reactor now looked more like an oil refinery or a steel works. With binoculars they looked down on the burning graphite, the red-hot biological shield, and the sinister blue glow in the core. It was terrible, but it was also awesome, and Legasov realized for the first time that they were facing a catastrophe of a historic kind, like the eruption of Vesuvius that had destroyed Pompeii, or the earthquake and fire in San Francisco.

The helicopter brought them down on the main square in Pripyat, and they went to the office of the party secretary to discuss what was to be done. While the others had been above the reactor, General Pikalov had made a survey of his own in an EMR2, a tracked reconnaissance vehicle based on the T54 tank for which Pikalov himself had drawn up the specifications. It was hermetically sealed and fitted with filters that removed radionuclides from the outside air. External dosimeters measured radiation, and lead panels in the armor protected those inside from gamma radiation. Three of these vehicles had been sent from Kiev. Pikalov had taken the first, and with a junior officer as his driver had set off from party headquarters in Pripyat, crashed through the wire fence that surrounded the power station and driven up to the wall of the damaged reactor. He was able to measure the level of radiation, and also the rate of emission of short-lived isotopes to help determine whether or not the reactor still remained active. This reconnaissance took about thirty minutes, and despite the protection of the armor gave the general and his driver a dose of thirteen rems.

With the data Pikalov had collected, together with measurements taken by Legasov and Abagyan, it was now clear that fission had ceased, but the burning graphite was still spewing millions of curies of radioactive particles into the sky. Something had to be done to put it out, and now Shasharin and Poloshkin, Marin and Mayorets, Ivanov and Pikalov, as well as Scherbina himself put forward a number of ideas—some wild and fanciful, like lifting huge tanks of water by helicopter and dropping them into the inferno, or building concrete walls around the reactor so that the fire fighters could approach it and spray it with water and foam. All were determined to take decisive, heroic action, but no one had the slightest idea what it should be.

As he listened, Legasov was dismayed by the ignorance of the officials and engineers. The operators had behaved well, remaining at their posts and obeying any orders they had been given, but those who had the scientific and technical knowledge to understand what had happened were paralyzed with indecision, and those who had the authority and the habit of command knew nothing about atomic physics. Only one man combined both qualities, and that was Legasov. He now described the situation to Scherbina as simply as he could. Fission had ceased in the reactor, but the graphite in the core was burning, sending millions of curies of radionuclides into the air. The average rate of combustion of graphite was about one ton per hour, and unit 4 had contained twenty-five hundred tons. Some of this had been thrown out by the explosion, but if as much as half remained it would mean that the fire would go on burning for nearly two months. It was therefore unacceptable to wait for the fire to burn itself out; clearly it must be extinguished.

There was a further and possibly more serious danger. If the temperature in the core increased, there was a danger that the uranium fuel itself would melt, with unforeseeable consequences. It was therefore imperative that the temperature in the core be reduced, but this could hardly be done with water; indeed, at temperatures over $2,500°C$ water itself breaks up into the explosive chemical components of hydrogen and oxygen. The only method to stop the emission of radionuclides and put out the fire was to smother the burning reactor with sand. This would have to be dropped into the crater by helicopters, and to the loads of sand should be added boron, dolomite and lead. The lead would cool the core; it boiled at $1,744°C$ and would absorb some of the heat. At these high temperatures the dolomite would break down into magnesium, calcium and carbon dioxide, which would also absorb the heat and generate an inert gas to smother the fire. Finally, the boron would absorb the neutrons and inhibit any further chain reactions.

Now at last Scherbina could show what he was made of; his dynamism could be released. He summoned the local air force commander, then turned to the assembled experts. How much sand would be required? There was a hurried discussion. The deputy minister of energy, Sasharin, estimated that the crater was four meters deep and twenty meters wide; therefore it would require three or four thousand tons of sand to achieve their objective. Was sand available? There were large deposits on the banks of the Pripyat River. And boron, dolomite and lead? Brukhanov and Kizima were told to make an inventory of what was available; otherwise supplies must be commandeered from farther afield.

The young air force commander, Major General Antoshkin, now reported to Scherbina, having arrived in Pripyat an hour before, at 1:00 A.M. on the 27th. Scherbina described his mission: he was to seal the crater of the reactor with sand. "Everything now depends on you and your pilots, Comrade General. Only you can save the day."

"And when should the operation commence, Comrade Minister?" asked Antoshkin.

Scherbina looked astonished. "Why, at once."

3

Antoshkin had never been to Chernobyl before and had not yet seen the reactor, having arrived by car in the dark. It was still dark, and the only helicopters at his disposal equipped for night flying were two Mi-6s at the Chernigov airfield. He would have to establish a flight path, a landing pad and a control tower. But where? He asked for advice. From where could he get the best view of the power station? Someone suggested the Hotel Polessia next door. Antoshkin crossed to the hotel from the party headquarters and went up onto the roof. From there he looked out toward the glowing reactor. Then he looked down to the square below. That could be the landing pad, and this would be the control tower. He went back to the party headquarters and proposed this to Scherbina, who grumbled that the noise would interfere with the work of the commission. But Antoshkin insisted, and the minister agreed.

At 4:00 A.M. Scherbina and the members of his commission finally retired to snatch some rest in the Hotel Polessia, only to be awoken two hours later by the roar of the huge Mi-8 helicopters that Antoshkin had ordered up from Chernigov at first light. The first, piloted by a Colonel Serebryakov, immediately set off on a reconnaissance mission over the reactor.

Meanwhile Antoshkin asked Scherbina where he could obtain the sand. With his natural impatience turned to exasperation by a short night's sleep and the sudden appearance of a dry throat and irritating cough, Scherbina lambasted the air force commander in true Soviet style. Why turn to him? Had he not men under his command? Tell *them* to dig up the blasted sand! His men were pilots, Antoshkin replied. If they had to dig up the sand, their hands might shake as they flew over the reactor. Scherbina looked around him furiously and saw two deputy ministers, Shasharin from the Ministry of Energy and Meshkov from the Ministry of Medium Machine Building. "Here, General," he said. "Here are two comrades who will do the loading for you. Find some shovels and bags and sand. There must be plenty of sand around; the whole place is built on it. Find some, dig it up and drop it on the damned reactor!"

There was no arguing with a deputy prime minister. Meshkov, Shasharin and Antoshkin left the party headquarters to look for sand. They walked toward the river and there, next to a café, they found some that had been dredged out of the river for the construction of a nearby block of flats. Sacks were brought from a warehouse, and finally shovels, whereupon the general and the two deputy ministers set to work to fill the bags with sand.

Back at the party headquarters, Scherbina called for Kizima to summon a brigade of his construction workers, but the men were not to be found. Therefore he turned to three experts who were standing by, the manager of one of the construction combines for the atomic industry, Nikolai Antonshchuk, and his chief engineer, Anatoly Zayat, along with the head of the country's hydroelectric operations, and ordered them to help. The three men ran down to the river. The sun had risen, and it was already hot; they found Meshkov, Shasharin and Antoshkin bathed in sweat. As they worked, Antonshchuk asked Shasharin whether the Ministry of Energy would pay out bonuses for those engaged in hazardous work. Shasharin, who already realized how dangerous it was simply to be in Pripyat, wearily said yes.

It quickly became apparent that six men were not going to be able to bag three thousand tons of sand, so rather than wait for Kizima and his men, Antonshchuk and Zayat drove off to the nearby Friendship collective farm. On this sunny Sunday morning, some of the peasants were working in the fields, others digging and planting on their private plots of land. Suddenly two sweating executives drove up and told them that they were wasting their time, that the reactor was damaged, that the land was contaminated, and that their help was needed to smother the reactor with sand. The peasants laughed at them and went on with their work. The two men drove on to a second group working in the fields. Again they told the same story and again they were ignored. They drove to the village to find the director of the farm and the party secretary. Slowly the truth sank in, and soon one hundred men and women from the Friendship farm drove to Pripyat to fill bags with sand.

Meanwhile, Colonel Serebryakov had returned from his reconnaissance mission over the reactor. The plan was to drop the sand from as low an altitude as possible to reduce the clouds of radioactive dust that it would raise. The colonel was also warned that the work would have to be done quickly; at a level of 250 meters the level of radioactivity was three hundred rems per hour.

Serebryakov reported to Antoshkin that the lowest they could safely fly over the reactor was two hundred meters; any lower and the hot air would foul the engines. The approach should be made at a speed of fifty kilometers an hour. The first batch of bags was brought up from the river and loaded onto the helicopter under the lash of Scherbina's tongue. Shouting over the roar of the helicopter's engines, he commented acidly on how good they were at blowing up power stations and how lousy at loading sand. Finally the the helicopter took off with Colonel Serebryakov at the controls. Guided by a ground controller on the roof of the hotel and by his navigator, on board the helicopter, he hovered over the target while two members of the crew opened the door. As a blast of hot air hit their faces, they lowered the six sacks of sand over the side. It took seven long seconds for them to fall, but they finally disappeared with a puff into the white eye of the crater.

As the day wore on, Colonel Serebryakov and his crew flew twenty-two missions over the reactor; another twenty-two were flown by a second team under Lieutenant Colonel Yakovlev. By 4:00 P.M. a new squadron had arrived in Chernigov from Torzhuk. At 7:00 P.M. Antoshkin and Shasharin reported to Scherbina with some satisfaction that 150 tons of sand, as well as some boron, dolomite and lead, had been dropped on the reactor. Scherbina was furious. What was 150 tons? A drop in the ocean! They would have to do better than that.

All that afternoon, while the helicopters were flying back and forth from Pripyat to the reactor, the long line of six hundred buses moved through the town to evacuate the population.

When the commission had reconvened at 7:00 A.M. on 27 April it had again received conflicting advice. Pikalov reported that there had been a slight rise in the levels of radiation in the town, but that there was little danger as yet of anyone receiving a dose of twenty-five rems. However, contamination north of the reactor was much higher, so there would be some danger if the wind changed. Some argued that since the norms existed, they should stick to them; others that while the whole-body dose might be well within the limits, the *actual* dose received by the thyroid would be many times greater, and that Pripyat was a young town with seventeen thousand children. In such a volatile situation, it was better to be safe than sorry. Scherbina was persuaded. At 10:00 A.M. he ordered General Berdov to proceed with the evacuation.

4

An energetic, good-looking, dark-haired man of medium height whose whole life had been spent in the militia, Berdov now put into effect the plan he had drawn up the day before. He faced a considerable task: to move fifty thousand people in a matter of hours, with minimum exposure to the open air. He had read in a Soviet newspaper that after the accident at Three Mile Island, the evacuation had taken five days and that eighty-seven people had died in the rush. He was determined not to make the same mistakes. Therefore he drew up a schedule for the buses to pick up people from their homes so that no one would have to wait in the open air, and to move them out in a coordinated convoy before dispersing to the neighboring towns and villages that he had alerted to prepare for the refugees.

Already that morning, members of the Komsomol Youth Movement had gone from block to block in Pripyat handing out iodine tablets for the children and advising people to stay indoors. Some took their advice; others did not. It was such a beautiful day that it was difficult to confine children to stuffy flats. However, at midday the announcement was made over loudspeakers that the town would be evacuated. Everyone was to return to his or her block of flats and to prepare for an absence of three days. A guard was put at the entrance of every block, and no one was to be allowed to clog the roads by leaving in their own cars.

Some of the families had been forewarned. The taciturn Vadim Grishenka had already told his wife, Ylena, to pack a suitcase because evacuation was inevitable.

"For how long?" she asked.

"Probably forever."

"But what about the May Day parade?"

"There won't be a May Day parade."

Not knowing when she would return, or how secure the flat would be in their absence, Ylena put together all the things she prized most: her leather coat, her fur coat, her gold bracelets, her Italian shoes and her ice skates. Her daughter had a new rucksack, which she filled mostly with her party dress. By contrast, Lubov Lelechenko, assuming that they were to be taken out into the surrounding forest, put on her oldest clothes. Luba Akimov, who had spent the morning trying to keep her children away from the windows, along with Inze Davletbayev, Natasha Yuvchenko and the other wives of the injured operators who had been flown to Moscow, obediently packed their bags and waited for the buses.

Katya Litovsky, the pretty girl whom Nikolai Steinberg had recruited into the turbine hall in the early days of the Chernobyl power station, had to prepare not just her ten-year-old daughter, but her daughter's friend, who was visiting. Lubov Lelechenko got a call from her daughter in Kiev, who wanted to know if they were going to Poltava for the May Day holiday, as planned, or should she come to Pripyat? Lubov dared not tell her daughter what had happened, so she handed the telephone to her husband. "We won't be going to Poltava," he said. "But don't come here."

The editor of *Tribuna Energetica*, Lubov Kovalevskaya, told her mother that she was sure she would never return. Like Ylena Grishenka, she packed her best clothes, her party dresses and an expensive woollen shawl. She told her daughter and her niece who was staying at the time to change into clean clothes, and while they did so she packed their rucksacks, but failed to notice, as they went down to the entrance of their block of flats, that her old mother was still wearing her slippers.

One by one, the families climbed into buses, and when each was ready the policeman in charge reported by radio to the central controller. The numbers were fewer than expected: from a population of about fifty thousand only twenty-one thousand were loaded onto the buses and trains—an indication of how many had already fled. Flats were checked to make sure they were empty; and finally the signal was given for the convoy to move out.

After only two hours, General Berdov could report to Scherbina that the operation was complete. It was a remarkable achievement, a tribute to Berdov's abilities and to the docility of the people. There was no hysteria and few complaints. But looking out of the window of the bus at the stricken power station, Lubov Lelechenko remembered the sense of doom she had felt when she had first caught sight of it on the hydrofoil coming from Kiev.

At Ivankov, forty miles from Chernobyl, the convoy split up, some buses going southeast toward Kiev, others northwest toward the towns of Narodici and Polesskoe. Katya Litovsky, with her daughter and her daughter's friend, was set down in a village three miles from Polesskoe. The director of the local collective farm, responsible for finding the evacuees somewhere to stay, invited the peasant women who had surrounded them to make their choices. It was like a cattle market, the peasants with their gnarled faces, gold-capped teeth and tight headscarves staring at the crowd of urban folks in their best clothes. An old babushka invited Katya Litovsky and the girls back to her cottage, muttering as she went that it was just like the war. When they reached her cottage, they were given some food and then retired for the night, the three of them in one bed.

The cheerful Ylena Grishenka in her elegant leather coat was given a friendly welcome by a family of poor peasants in the village of Pukhovka, where she shared a bed with her daughter. There was no bathroom, but the next day some soldiers came to the village and rigged up an outside shower with water pumped from the pond. Everyone assumed that they were outside the zone of contamination. When their hosts asked Ylena if she would like to milk the cows, she said she was frightened of them. Nor could she help dig up potatoes; she had brought only her jewelry and Italian shoes.

Lubov Lelechenko, who had come with only a single change of clothes, was billeted in the house of the local collective-farm director in a village near Ivankov, together with the wife and daughter of Lyutov, the station's scientific deputy chief engineer. The wife of the collective's director was a teacher like Lubov, and the next morning took her to the village council because the teachers from Pripyat had been told to continue with their classes. But when Lubov stood up in front of the children and tried to teach them mathematics, she found that she was so disoriented she could not add the simplest sums, and her pupils had to help her.

Further dispersal of the evacuees was strictly controlled; members of the Communist party had their cards confiscated "for safekeeping." Many left anyway, particularly those who, like Tatiana Palarmachuk or Katya Litovsky, had families in the Ukraine. It was difficult to get permits or money, and some parents lost their children in the confusion. Many people had a dry throat, some a hacking cough, and all were drawn and exhausted after being torn so abruptly from their beautiful town and comfortable homes. The very backwardness and isolation of the villages of the Polessia made them realize how agreeable life in Pripyat had been. As the old peasants kept repeating, it was just like the war.

5

That Monday morning, 28 April, a thousand miles to the north, the alarm went off at the Fosmark nuclear power station one hundred kilometers from Stockholm when a worker passed through the dosimetric control at the end of his shift. The levels of radioactive contamination on his clothes greatly exceeded the norm. Fearing that there was a leak from the Swedish reactor, evacuation was ordered of all inessential personnel.

Before the leak could be traced, meteorological stations in other parts of Sweden reported radioactive particles in the wind blowing across the Baltic from the Soviet Union. At first it was assumed that despite the test-ban treaty there had been an experimental detonation of a nuclear weapon, but an analysis of the isotopes soon established that the radionuclides must have come from a nuclear reactor. Although five times higher than background level, this radioactive contamination was not thought hazardous for the Swedish population, but it did suggest that there had been some kind of accident, possibly in the Ignalina plant in

Lithuania. Swedish diplomats were instructed to make urgent inquiries to Moscow. Calls were made to three government agencies, including the Atomic Energy Commission; all denied that an accident had taken place. It was only at 9:00 P.M. that a statement was issued by the Soviet government's official news agency, TASS:

> An accident has occurred at the Chernobyl Atomic Power Plant. One of the atomic reactors has been damaged. Measures are being undertaken to eliminate the consequences of the accident. Aid is being given to to those affected. A government commission has been set up.

The same statement was broadcast on the Moscow television news program "Vremya." It was the seventh item of news.

6

The brief admission that there had been an accident at Chernobyl had been agreed to after a long debate at a meeting of the Politburo that Monday morning. The news that fallout had been detected in Sweden told the Soviet leaders what their own subordinates had tried to conceal—that the explosion in the fourth reactor was not an industrial accident but a major disaster. The entire resources of the nation would have to be mobilized to deal with the catastrophe, and to do this a second commission was formed within the Politburo. It was headed by the prime minister himself, Nikolai Ryzhkov, and included among its members Yegor Ligachev, Vitali Vorotnikov, and the ministers of the interior and defense.

The Soviet leaders were acutely aware of the political as well as the ecological consequences of the disaster. Not only the accident but the delay in announcing it had damaged their country's prestige abroad. It had been humiliating to have had the truth forced out of them by Swedish diplomats. To be able to counter the malevolent exaggerations of Western propaganda, it was vital to have precise data from Soviet sources.

Measurements within ten kilometers of the reactor were supplied by

General Pikalov. Already, on 29 April, he had sent in three groups of specially equipped helicopters, staggered at heights from fifty meters to two kilometers. Every two minutes each helicopter sucked in a sample of the air. By measuring the radioactivity, Pikalov was able to form an accurate picture of the level of contamination, and he reported to Ryzhkov and Scherbina that the active emission every twenty-four hours was about one hundred times higher than that estimated by the physicists on the ground.

To measure contamination over a wider area, Ryzhkov summoned the chief of the State Committee of Hydrometeorology, Yuri Israel. A tall man with a burly Russian look that belied the Semitic connotations of his family name, Israel had been a student of the eminent geophysicist Fedorov, and later his deputy director at the Institute of Hydrometerology. He had considerable experience in measuring the levels of radioactivity in the atmosphere; at the time of the test-ban treaty, he and his colleagues had set up four thousand observation stations on land, in the air and on satellites in space to measure the different parameters in the earth's atmosphere.

Alerted by Ryzhkov on 29 April, Israel assembled the same group of specialists and set off for Kiev. Eight airplanes and helicopters equipped with measuring equipment took off to track the plume of short-lived isotopes emanating from the reactor, while teams set out to measure ground contamination by the long-lived isotopes like cesium 137, strontium 90 and plutonium 239. At first it was difficult to distinguish between the radioactivity in the air and on the surface; but still, they were able to produce rough maps to show the contaminated areas by 1 May.

The man called upon to advise the commission on what to make of this information was the director of the Institute of Biophysics, Academician Leonid Ilyn. The son of an engineer and the grandson of an engine driver, Ilyn had served five years in the navy as a medical officer attached to the Black Sea fleet after graduating from the Leningrad Medical Institute. In 1963 he had returned to Leningrad to do research at the Institute of Radiological Hygiene. It was from here that he had traveled to Mayak to study the aftereffects of that accident, and to the nuclear testing grounds in Kazakhstan. In a safe in his office, Ilyn kept a lead box that held a wooden Russian doll containing the little black pearls formed by the melting of sand at the epicenter of a nuclear explosion.

The field of Ilyn's research was the protection of human beings from radioactivity, and frequently he and his colleagues would conduct experiments on themselves to test the efficacy of stable iodine against contamination by radioactive iodine 131. He was also an able administrator; at the age of only thirty-three, he was appointed a deputy director of the Institute of Radiological Hygiene, and in 1967, before he was forty, he was chosen by Slavsky, the minister of medium machine building, to head the ministry's Institute of Biophysics in Moscow.

A man of medium height, with strong features and a robust voice, Ilyn had kept the commanding manner of a naval officer. It was not difficult to imagine him as a provincial governor in czarist times. He was a member of the Communist party, but he was less of an ideological zealot than a Russian patriot, proud to have been born in Smolensk, where the armies of both Napoleon and Hitler had been routed; proud, too, of the achievements of Tolstoy, Tchaikovsky and Lenin, who had thrust his backward nation into the vanguard of mankind. A huge portrait of Lenin in inlaid wood hung on the wall of his office, which looked out over the Moscow River.

For all his qualities as a leader, Ilyn remained a scientist, and he continued his research into the effects of radiation. His work was classified, but his achievements were recognized, and in 1974 he was elected to the Soviet Academy of Medical Sciences. As his institute was funded by the rich Ministry of Medium Machine Building, much of its research related to defense—in particular, the provision of data on the possible consequences of a nuclear war. In 1970, soon after he had become director of the institute, Ilyn had co-written and published a circular entitled "Temporary Instructions on Measures to Be Taken to Protect the Population in the Event of Nuclear Explosions." The circular's main point was the need to protect the thyroid gland from radioactive iodine 131 by taking pills of stable iodine. It also envisaged a serious leak of radioactivity from a civilian reactor, recommending a detailed contingency plan at every nuclear power station.

It was this document that specified the norms for evacuation, as well as recommending doses of iodine. It was approved by the deputy chief medical officer of the Soviet Union, but only a thousand copies were printed. It was updated in 1984 under the Commission of Radiation Safety, headed by Academician Ilyn, with only some minor amendments, but was given no wider circulation. To prepare too well for an accident might lead people to believe that it could happen.

More plausible was a war, and here Ilyn had been drawn into the propaganda battle with NATO strategists in the West. Because of the Soviet advantage in conventionally armed forces, it was said that the West would have to defend itself with nuclear weapons, using short-range rockets like the Minutemen and Cruise missiles against the Soviets' wave of tanks.

Inevitably, the advantage of Soviet conventional superiority would be retained if public opinion in the West could be convinced that a nuclear war of any kind would have disastrous consequences for both sides. In March 1981, with Brezhnev's personal physician, the cardiologist Yevgeni Chazov, later the minister of health, Ilyn attended the First Congress of International Physicians for the Prevention of Nuclear War, in Airlie, Virginia. In 1984, with Chazov and Angelina Guskova, he published a book in both Russian and English called *Nuclear War: The Medical and Biological Consequences: Soviet Physicians' Viewpoint.* Although peppered with Cold War rhetoric ("There is truly no limit to the cynicism of top Washington officials and their NATO accomplices"), essentially it was an extrapolation of the research done on survivors of Hiroshima and Nagasaki, which showed the extensive and long-lasting harm done by radiation.

The intention of the book was to alert the public to the catastrophic aftermath of a nuclear war, and so to put pressure on Western strategists to abandon their "cynical" first-strike option. Therefore it made the most of Japanese research, which established "a greater number of identifications of certain types of cancer" among those who received "a more than 100-rad irradiation dose":

> Leukosis, especially in its acute forms, is one of the diseases that has been proved beyond reasonable doubt to stem from nuclear explosion effects. Leukemic transformations of hemopoiesis are most likely to occur within 5 to 10 years following irradiation resulting in that period in a greater (fivefold to tenfold) increase of leukemia, as against reference groups of persons who received a less than 10-rad dose. The United Nations Scientific Committee on the Effects of Atomic Radiation publications over a number of years also contain strong evidence of excessive mortality from leukosis among men and women of all ages who as a result of nuclear

explosion received doses of 50–200 rads, as compared with groups with less than a 10-rad dose and nationwide statistics.

For those who received a dose of more than two hundred rads, the probability of developing other forms of cancer was doubled; only for uterine and pancreatic cancer was it the same, and for rectal cancer the risk was reduced. Nor were smaller doses necessarily harmless; in his invective against the neutron bomb, Ilyn warned that "even small radiation doses will affect people very seriously." A dose of fifteen rads would not lead to radiation sickness, but "subsequently some of those irradiated are likely to develop malignant tumors or leukemia. Negative genetic consequences may occur in several generations of the descendants of those initially exposed."

VI

1

On 29 April, on Ryzhkov's instructions, Academician Leonid Ilyn flew
to Kiev. That very day, the direction of the wind changed. Instead of
blowing from the southeast, carrying the millions of curies of radioactive
isotopes still spewing out of Chernobyl toward the relatively sparsely
populated territory to the north, it now carried them back across the
Ukraine toward its capital city, Kiev.

Upon landing at Juliana airport, Ilyn was immediately besieged by
anxious officials desperate to know what danger they faced. This filled
him with a sense of foreboding; suddenly he realized how few people
were qualified to assess the risk. In the entire Soviet Union, the number
of real experts could be counted on the fingers of one hand. The level
of radioactivity in the city of Kiev was 2.2 millirems per hour; if it
remained at this level, it would take more than a year to reach the
minimum dose of twenty-five rems. However, he knew that even a small
dose received by a large number of people could have serious consequences for a few. There was no absolute safety but only an element of
risk that had to be weighed in the balance, not so much by scientists as
by political leaders.

The city was in full bloom. Built on a bluff overlooking the Dnieper
River, with the golden domes of its many monasteries glinting in the
sun, it reminded visiting Muscovites that it had been the first capital of
the Christian kingdom of Rus. Lying in the midst of rich farmland
stretching to the Black Sea, it had for a millennium been the trading

center for the exchange of exotic fruits from the warmer climate of the southern Ukraine for furs and timber from the forests to the north. Despite the destruction wrought by war and revolution, and the heavy monuments to socialism and the Motherland that had been built in parts of the city, the undulating, tree-lined boulevards and the great curve of the wide Khreshchatyk still suggested a lightness and vitality that contrasted with the dour thoroughfares of Moscow and Minsk. It was a city for spring, when all at once, throughout the Ukraine, the fruit trees came into blossom and the long gray winter was transformed into a colorful, fragrant world. This particular spring was the finest that any could remember. The sun was as hot as in midsummer, and the whole city of three million people was in a state of high excitement as it prepared for the festivities surrounding the May Day parade—a celebration of the triumphs of socialism, of course, but also an improvised festival that served as an excuse to drink up stocks of vodka before the new laws against alcoholism came into force.

Everyone was busy. Carpenters were building the dais on the Khreshchatyk from which the party leaders would watch the parade; municipal workers were winching up the huge portraits of Marx, Engels and Lenin; young Komsomol activists raised the red bunting and banners with the slogan PROLETARIANS OF ALL COUNTRIES, UNITE!

The journalists prepared their copy for the May Day editions of their papers:

> Soviet people are celebrating this May Day—the day of the working people's international solidarity—as a drastic turning point in the dynamics of the substantial acceleration of the country's development. The emotional content of the Twenty-seventh Congress of the Communist Party of the Soviet Union—a congress of Party principledness, exactingness and Bolshevik truth, the optimism and businesslike nature of its life-giving ideas, the grandeur and realism of the plans it put forward—that is what is the all-embracing determining basis for the sociopolitical and moral atmosphere in which Soviet society is now living and which also dominates today's celebrations.

To link the wonderful weather to the promise of *perestroika* was irresistible:

Each May Day has its peculiar distinctive features. This year it had a powerful ideopolitical and revolutionary incentive: the 27th CPSU Congress was held and it opened for the Soviet people a path to new accomplishments, to new qualities of life, to an accelerated socioeconomic development of our society.

But international ramifications were not to be forgotten:

An important feature of this May Day, too, was the fact that it marked a jubilee. A hundred years have passed since the day when the authorities cruelly suppressed a general strike by the American workers, a day that initiated the celebration of the day of the Working People's International Solidarity. The red banners waving in the May breeze seemed to converse with the red calico of slogans and posters. By the side of the Central Tribune fly the state flags of the USSR and all Soviet republics, in whose family the Soviet Ukraine has also won its happiness.

Presiding over the celebrations was Vladimir Shcherbitsky, the general secretary of the Communist Party of the Ukraine. One of the last survivors of the Brezhnev era, he had been obliged to dismiss nine of his twenty-five regional secretaries during Andropov's drive against corruption, though he himself had survived.

Alerted to the accident at Chernobyl, the various forces at the disposal of the Ukrainian government had been dispatched with exemplary speed and skill. The militia, the civil defense and the troops of the Ministry of the Interior and the KGB had sealed off the stricken area before sunrise on 26 April. No one had worn protective clothing of any kind.

Less alert were the officials of the Ukrainian Ministry of Health, who, like their counterparts in Belorussia and the Russian Federation, were caught totally unprepared by the accident. Assuming, like everyone else, that Soviet nuclear reactors were safe, they had made no contingency plans for mass protection against radiation. Moreover, the power station and the city of Pripyat were outside Ukrainian jurisdiction. Chernobyl was an All-Union facility whose medical services were the responsibility of the Third Division of the Soviet Ministry of Health.

An added misfortune was the absence at the time of the accident of the Ukrainian Minister of Health, Anatoli Romanenko; he was at a conference in Atlanta, Georgia. When the scale of the catastrophe became apparent, his officials prepared to cater to the evacuees from Pripyat and the ten-kilometer zone. Up to one hundred teams were formed to take care of them, but treatment was circumscribed by the available resources. The supplies of stable iodine were quite inadequate for it to be distributed to the whole population, and there were no clear guidelines as to whether or not it was required, or what threat was posed by the escaping radiation.

While many senior figures in the party and the administration with private knowledge of the gravity of the accident at Chernobyl sent their children and grandchildren out of the city, the Ukrainian leaders accepted Ilyn's judgment that the people of Kiev were not at risk, and preparations continued for the May Day parade.

2

Ilyn now flew north to Chernobyl and had the plane circle above the fourth unit. Once again, as with Scherbina, Legasov and Pikalov, it took the sight of the destroyed reactor to make him appreciate the magnitude of the disaster. As the plane flew over the deserted town of Pripyat, he noticed a line of diapers strung across a balcony on the top floor of a nine-story block of flats, and just above, on the roof, a red and black banner proclaiming, "LONG LIVE 1 MAY!" The sight brought home to him that this industrial catastrophe was also a human tragedy. But his faith in the Soviet system was not shaken. When he arrived at Party Headquarters and saw the resources that had been marshaled to contain the accident, he felt that it could only have been done so rapidly and effectively under a centralized, authoritarian form of government.

Everywhere there were signs of vigorous activity. The decision had been made that day by Scherbina and the commission to move its headquarters from Pripyat to the town of Chernobyl, twenty kilometers from the power station. The hospital, too, was to be evacuated and the contaminated town abandoned. A holiday camp for young Pioneers—the Boy Scouts of the Communist party—had been requisitioned to provide living quarters for the doctors, nurses and power station personnel. Pikalov's troops were housed under canvas in uncontaminated territory, sixty to a tent; others were billeted as far away as Ivankov, sixty kilometers from Chernobyl.

The bombing of the reactor with sandbags had now reached a rate of over 180 runs a day. The drivers of the Atomic Energy Transport Association had been organized to dig up clay from a quarry near the village of Chistogalovka, only five kilometers from the power station, and to deliver it in sacks to the helicopter pads. Major General Antoshkin, the air force commander, now had at his disposal the heavier Mi-6 and Mi-26 helicopters, which had flown in from Torzhok. The pilots, many of them veterans of the war in Afghanistan, were required to exercise great skill in horrific conditions. Guided by the controller on the roof of the Hotel Polessia, they had to fly low enough to hit their target, yet avoid the many pylons and cables close to the power station and the tall chimney that rose from the fourth unit.

It was very hot, but they dared not switch on the fans to ventilate the cabins for fear of drawing in radioactive dust, so they sweated profusely with both heat and tension. They kept the crews to a minimum—usually no more than two pilots and a navigator. Some rudimentary measures were taken to protect them. Before setting off from Chernigov airport at 4:00 A.M., they were given iodine pills with their breakfast, and they were instructed to take more when they returned at 11:00 P.M., but most were too exhausted to remember. Some sheets of lead were placed beneath the seats of the helicopters, but any added weight detracted from the load each helicopter could carry.

The airmen carried dosimeters and were withdrawn from duty when they had received a dose of twenty-five rems. However, because of the war in Afghanistan, there was a limited number of people available. Knowing the danger, many were afraid. Party zealots among the pilots formed a group to raise morale, using the slogan "Any effort, energy and, if necessary, one's life to fulfill the government's tasks!," but it did not always work. A young pilot from the Torzhok squadron at first refused to fly but was finally persuaded by a companion that the job had to be done. Others summoned up Dutch courage, calming their nerves when they returned at night by drinking the alcohol stored on board the helicopters as a defreezing agent. It helped them sleep and was still on their breath in the morning. Their commanders smelled it but made no objection.

In the first days Scherbina complained to Pikalov and Ivanov that

Antoshkin was incompetent, but subsequently he recognized the efficacy of the calm young commander's operation. On 28 April Antoshkin's helicopters made 93 sorties over the reactor; on the 29th this had increased to 186, and the payloads had grown larger by using upturned parachutes as nets to hold between six and eight sacks at a time. On the 30th they estimated that they had dropped over a thousand tons of sand, clay, boron and lead. Moreover, the strategy seemed to be succeeding: the emission of radionuclides was decreasing from an estimated 12 million curies on the day of the accident to 4 million on the 27th, 3.75 on the 28th, 3 million on the 29th and little over 2 million on 30 April. On the eve of May Day, Valeri Legasov could feel satisifed that he had given Scherbina the right advice.

3

Not everyone agreed. On 27 April two experts on the RBMK reactors from the Kurchatov Institute had arrived at Chernobyl. Like Rosencrantz and Guildenstern to Legasov's Hamlet, Alexander Kalugin and Konstantin Fedulenko had been brought from Kiev's Juliana airport in a minibus carrying nine or ten scientists, their measuring equipment and a special camera from NIKYET. Beyond Ivankov, they were held up by the long line of buses coming from Pripyat, and only reached their destinations in the evening.

Checking into the Hotel Polessia, they were asked to pay two rubles in advance for their room. The hotel's restaurant was closed, so they went out into the streets to look for some supper. They came to a café surrounded by a crowd of people; everything in it was being given away free. All the food was gone, but there were packs of cigarettes, even the coveted BTs (Bulgarian Tobacco), which were normally hard to find. The scientists filled their pockets and sauntered on. Some people were in uniform, others in civilian clothes. A dosimetrist they knew advised them to remain indoors, and so, despairing of supper, they returned to their rooms in the hotel, where, on the advice of the dosimetrist, they moved the beds away from the window.

Waking hungry on the morning of the 28th, the scientists went to the headquarters of the commission in Party Headquarters, where they were given a breakfast of sausages, white buns, lemonade and iodine tablets. When Fedulenko asked whether they should not have taken the iodine tablets earlier, he was told "better late than never."

They reported to Legasov. Their task, he said, was to discover how the accident had occurred. The computer printouts and tape recordings were waiting for them in the bunker beneath the administrative block. He took two dosimeters out of his pocket and handed one to each of the two scientists.

They took a bus to the power station. It was shrouded in fog, a natural mist mingling with the vapor that still rose from the ruins of the fourth unit. Both scientists—especially Kalugin, whose whole life had been dedicated to RBMK reactors—were appalled to see all the mangled pipes and tanks exposed to the open air. They got off at the administrative building in front of the first unit and went down to the bunker. There they were met by Alexander Gobov, the director of research at Chernobyl and an old friend of Fedulenko's. He gave them the logbooks from the fourth unit, printouts from the Skala computer registering one hundred parameters of data, and the tapes on which were recorded the voices of the operators as they had conducted the tests.

Looking first at the logbooks, Kalugin saw that the reactor had been put on to low power twenty-four hours before the accident, with only thirteen control rods inserted into the core—two fewer than the minimum required—and that this clear breach of the guidelines had been ratified by the signature of the head of the shift. Turning to the printout from the computer, he saw that one minute before the accident had occurred, almost all of the boron control rods had been withdrawn from the core. The power of the reactor had been at two hundred megawatts. Thirty-six seconds after the test on the turbines had started, it had risen to six hundred megawatts. The AZ safety button had been pressed, and two hundred rods had started their descent into the core, but three seconds later the power had increased by a factor of three and the pressure in the reactor had doubled. Thereafter, the lines indicating neutron power and steam drum pressure ran vertically to the top of the paper.

Could it have been, Kalugin wondered, that as the control rods descended they had displaced the water in the base of the reactor, leading to a surge of positive reactivity? With growing dismay, he remembered that two years earlier this possibility had been raised at a meeting with his colleages at Dollezhal's institute, NIKYET. They had agreed that it was a remote possibility, but only if too many control rods were withdrawn. With the minimum of fifteen stipulated, such an uncontrollable leap in power could not occur.

Kalugin and Fedulenko then listened to the recording of the operators' voices. At times they were muffled and indistinct, but the scientists could make out an agitated shout of "Press the button." Clearly the operators had been aware of a crisis. Next Gobov showed them photographs taken on the morning of the 26th, which showed not just the destroyed tanks, exposed pumps and piping, but the large blocks of graphite that had been thrown out by the explosion. It was incontrovertible evidence that the reactor had been destroyed. Gobov told them that in places the dosimetric readings were so high that fragments of the fuel itself must also have been ejected.

Kalugin and Fedulenko had seen enough to reach a preliminary conclusion. They telephoned Legasov, and soon after midday a telegram was sent to the Politburo in Moscow: "Cause of accident unruly and uncontrollable power surge in the reactor."

Summoned back to Pripyat by Legasov, Kalugin and Fedulenko waited for the bus in front of the administrative block. It was now early afternoon, and the morning's mist had been dispersed by the heat of the sun. They smoked and chatted while watching the helicopters make their bombing runs over unit 4. Several others were waiting for the same bus—some wearing military uniforms, others, like Kagulin and Fedulenko, dressed in civilian clothes. No one wore a mask or seemed worried about radiation. The damaged reactor was hidden from their view by the wall of the turbine hall, but Fedulenko noticed that every time a helicopter dropped a load of sacks into the reactor a dark gray cloud of smoke rose from behind the ventilation chimney. It floated almost to the top of the chimney; then the darker particles appeared to sink down to the earth again, while the lighter-colored vapor rose above the chimney and was carried off to the north by a current of air.

After waiting in the open for more than a quarter of an hour, the group was picked up by an old Lvovsky bus. It was filled with dust and drove slowly past the fourth unit. Fedulenko had in his pocket a cloth mask that he had brought with him from Moscow, and though no one else was wearing a mask or respirator of any kind, he took it out and held it over his mouth and nose. Slowly the bus chugged past the damaged reactor. Another load of sand and clay dropped into the crater; another cloud of dark dust rose into the air.

Back at Party Headquarters, Kalugin and Fedulenko found that Legasov was less interested in their hypothesis about the cause of the accident and more in how to deal with its disastrous results. He described to them the action being taken to and told them to calculate the likely effect upon the core. They returned to the hotel and went to work with a slide rule.

The next morning, the 29th, they were told that they were to move to the Pioneer camp. When they had been driven 3.5 kilometers from the power station, they got out to transfer to a less contaminated bus. It was an exceptionally beautiful day, and the air was fragrant with blossoms. It was almost impossible to believe that the new shoots of grass were already contaminated by radiation. While they were waiting, Fedulenko watched some peasants tilling their land with horse and plough, while others planted potatoes. No one had told them that they were wasting their time.

Anatoly Serotkin, the chief physicist from NIKYET, came up to Kalugin, his face expressing the utmost dejection. "Have you got a gun?" he asked.

"No. Why?"

"I could use one."

He reminded Kalugin of the meeting held two years before at which they had discussed the theoretical possibility of an accident of this kind. Shaking his head, Serotkin admitted how mistaken they had been to assume that the operators would observe the regulations. Had there not been reports about the lack of discipline in the nuclear power stations run by the Ministry of Energy? Had they not considered establishing commissions, run by their Ministry of Medium Machine Building to verify standards? Fedulenko was obliged to agree. He remembered the remark made by Fomin that it was as simple to run a reactor as a samovar.

As the second bus passed through Chernobyl, they saw the peasants who were being evacuated loading their belongings onto carts. Children

were weeping; their mothers were distraught. What a triumph for socialism! This quiet country town, which had sat on its bluff above the
Pripyat River since the twelfth century, with its wooden houses and
onion-domed church, was now to be abandoned, probably forever.
Seeing this and feeling, like Serotkin, partly to blame for this catastrophe
that had shattered the lives of so many innocent people, Kalugin wanted
to lie on the floor of the bus to hide his shame.

4

At 10:00 A.M. on 1 May in Kiev the parade started on the wide curving
boulevard, the Khreshchatyk. Brigades from the Pioneers, the Komsomol Youth Movement and sports teams, as well as garlanded young
men and women in national costume, prepared to march past the party
élite, who stood on the podium in front of Lenin's statue. Next to
General Secretary Vladimir Shcherbitsky stood a stern matron, Valentina
Shevchenko, a deputy prime minister of the Ukraine. Other government
and party leaders were drawn up in ranks around them, and on either
side of the podium were seated veterans from the Great Patriotic War,
and leading figures from the scientific and cultural world, as well as
industrial leaders and record-breaking production workers. There were
also honored guests from abroad, including the representative of a Polish
building combine that was building a hotel in Kiev. "On this festive
day," he told a reporter, "I would like to wish the Ukraine's working
people great success and clear skies over the planet."

Some of those who had not gone to the Khreshchatyk or had fled the
city watched the parade on television. Sergei Sklar, for example, remained at home not because he was worried about radiation but because
the wind now coming from the north made it colder than it had been
a few days before. He knew about the accident at Chernobyl; he had
heard the announcement on "Vremya" on 28 April, and had been told
by friends who listened to the Voice of America that it was perhaps more
serious than the authorities had admitted, but nothing had led him to
suppose that either he or his wife or his eighteen-month-old son, Sergei,
were in any danger. The day before, the little boy's grandmother had
taken him out for two or three hours in the open air, and seeing so many

people smiling and dancing on the Khreshchatyk, he felt it must be safe enough for Sergei to be playing outdoors.

Yet, studying the crowds on the screen, he saw the occasional anxious expression among the smiling faces. He noticed, too, that a particularly gloomy look had replaced the usually genial countenance of General Secretary Shcherbitsky, standing on the podium with his grandson, and that he kept looking at his watch.

Then the telephone rang. It was a friend who warned Sklar that it was perhaps not as safe as they had supposed. Little Sergei was called indoors.

5

At 1:00 P.M., after the parade in Red Square in Moscow, Nikolai Ryzhkov held a meeting of the Politburo's Chernobyl commission in the Kremlin. When reports were given on the measures taken in the wake of the accident, it was immediately apparent that the existing structures were not coping. The civil defense and the Ukrainian Health Ministry had been overwhelmed by the scale of the disaster. Urgent coded requests had been sent from the Belorussian Ministry of Health for radiological testing equipment and supplies of potassium iodide tablets. The decision was made to mobilize the army medical corps—Ryzhkov signed the decree—and to form a medical commission in Moscow under the auspices of the Ministry of Health, including Academician Ilyn and the celebrated hematologist Professor Andrei Vorobyov.

The commission also decided to form a new team to relieve Scherbina and Legasov, who were not only exhausted but had absorbed significant doses of radiation. Ryzhkov appointed another of his deputy prime ministers, Ivan Silayev, a man of considerable experience in the aircraft industry, and told him to be ready the next day to leave for Chernobyl. Two scientists from the Kurchatov Institute were chosen to replace Legasov: the second deputy director, his great rival, the atomic-fusion expert Academician Yevgeni Velikhov; and Professor Eugene Ryzantzev, the head of the department for atomic safety, the superior to Drs. Fedulenko and Kalugin.

Ryzhkov also heard that there were differences of opinion between the civil defense, the chemical troops and the Ministry of Health about whether further evacuation was necessary beyond the ten-kilometer exclusion zone. Other experts were therefore put on notice to fly to Chernobyl the next day. Finally, late in the afternoon, Ryzhkov decided that he himself, together with Yegor Ligachev, would fly to Chernobyl, and that evening he telephoned Gorbachev to tell him that the situation seemed sufficiently serious to warrant his personal attention.

As befitted their rank, Ryzhkov and Ligachev were met by the Ukrainian general secretary, Vladimir Shcherbitsky; a deputy prime minister; and the first secretary of the Communist party of the Kiev region, Grigori Revenko. They were led to a cavalcade of large black limousines, and driven at high speed and with a motorcycle escort to Party Headquarters in the town of Chernobyl, which since the evacuation of Pripyat on 29 April had been the command center of the government commission. There they met with the large number of soldiers, scientists and engineers already working on the aftermath of the accident; together with ministers, deputy ministers and officials from Soviet and Ukrainian departments of state, the group that gathered at 3:00 P.M. before Ryzhkov and Ligachev amounted to 250 people.

On behalf of Scherbina, Legasov made his report. More than five thousand tons of sand, boron and lead had now been dropped over the burning reactor. As a result, the emission of radionuclides from the damaged reactor continued to decline. It was now only a little over two million curies per day—a high rate, certainly, but sufficient to suggest that smothering the reactor with sand was having the desired effect.

Following this optimistic line, the minister of energy and electrication, Mayorets, addressed the visiting leaders. "Dear comrades, members of the Politburo, Nikolai Ivanovich and Yegor Kuzmich, I am able to tell you that we have prepared the schedule for the reconstruction of the fourth unit and can assure you that it will be put back into operation by autumn, or winter at the latest. I can also assure you that the fifth unit will be launched on schedule." These assurances were met with applause.

Not all were so sanguine. General Pikalov's assessment of the situation was grave. An area within fifteen kilometers of the power station and the station itself were both heavily contaminated and should be evacuated at once. There was also the danger that any rainfall would wash radioactive dust into the rivers, which fed the Dnieper and the Kiev reservoir.

Ligachev interrupted him. "And how long, Comrade General, will it take to clear up this contamination?"

"Up to seven years."

Ligachev exploded. "We'll give you seven months! And if you haven't done it by then, we'll relieve you of your party card!"

"Esteemed Yegor Kuzmich," said Pikalov. "If that is the situation, you needn't wait seven months to take my party card. You can have it now."

Pikalov's pessimistic evaluation was confirmed by the maps that had been prepared by Professor Israel's hydrometerological committee, showing radioactive contamination outside the ten-kilometer zone. These showed deposits of plutonium 239 and 240 within a thirty-kilometer zone, while surface contamination by cesium 137 of over forty curies per square kilometer had been found beyond the thirty-kilometer zone, and even in territory to the north of Gomel in Belorussia. Some argued however, that the population was still in no danger of crossing the control limit of twenty-five rems. Distribution of iodine tablets would protect the thyroid from iodine 131, which had a half-life of only eight days. Cesium 137 on the ground had a half-life of thirty years, but if sensible measures were taken to ensure a supply of uncontaminated milk, the additional dose the populace would receive from the cesium would still be well within the prescribed limit.

Psychological factors had to be taken into account. The benefits of escaping additional contamination might be outweighed by the stress inherent in uprooting people from their homes. It was estimated that 116,000 people were involved. It should also be noted that those who had already been evacuated from Pripyat and the villages in the ten-kilometer zone had gone to territory hardly less contaminated than that which they had left—for example, the land around Narodici and Polesskoe. Some of the evacuees who had been sent there might as well have stayed at home.

Others were in favor of evacuation. The deputy chief of the civil defense, General Ivanov, had already prepared a contingency plan. He estimated that some of the villagers in the area could accumulate doses of six to eight rems, significant for adults and serious for children, and during the briefing the day before had recommended the evacuation of nine villages to Scherbina. Scherbina had put off a decision, but when Ryzhkov now asked what preparation had been made, Ivanov sent a note to the rostrum saying that if cattle were left behind, the area could be evacuated within a couple of hours.

Confused by this conflicting advice from the experts, Ryzhkov conferred with Scherbina, Revenko and Ligachev. They weighed the expense and the strain on resources of uprooting tens of thousands more people from their homes against the chance that if they remained some might be harmed by the radiation. Finally a decision could no longer be postponed. Ryzhkov thumped his fist on the table. No more discussion! The town of Chernobyl and all the villages were to be evacuated: there would be a thirty-kilometer exclusion zone.

When Ryzhkov and Ligachev left the town of Chernobyl that evening, the first buses were already arriving to remove the population. Their big black Zils, however, were too contaminated to leave the ten-kilometer zone; the prime minister and the secretary of the Central Committee had to return in smaller cars.

6

Although Velikhov and Ryzantzev had been sent to replace him, Valeri Legasov did not leave with Scherbina, but summoned Kalugin and Fedulenko to an urgent meeting. The most recent measurements taken by the chemical troops showed that on the very day that they had reassured the government that the reactor was under control, the emission of radionuclides from the damaged reactor had doubled from two million to four million curies per day. Rather than cooling down, the reactor core was getting hotter.

This gave Fedulenko the courage to criticize the whole idea of smothering the reactor with sand. "In my opinion, Valeri Alekseyevich," he said to Legasov, "it's counterproductive. All that happens is that radioactive dust is thrown into the air."

"We heard a broadcast from Sweden," said Legasov, "that said that burning graphite should be smothered with sand."

"But uranium needs no oxygen for a chain reaction," said Fedulenko. "And the sand is unlikely to put out the graphite fire. We saw the video made by Poloshkin from the air. It shows that the core of the reactor is still largely blocked by the biological shield."

"It was blown aside by the blast . . ."

"It has been moved, certainly, but it has only uncovered about twenty percent of the opening to the active zone. It's a narrow crescent shape, and the chances of the sand and boron landing inside are small. It will all end up on the shield."

"That doesn't matter. We'll bury the whole reactor in sand."

"But every load of sand throws heavy particles into the air. If you left it alone, they would stay in the reactor."

Legasov thought for a moment, then said, "But what will people think if we just leave it alone? We have to do something."

"And why the lead?" asked Kalugin.

"The lead melts and boils and removes heat from the core."

"But lead also poisons the atmosphere."

"That's less dangerous than a chain reaction."

"In my opinion there's no danger of that."

"So what do you suggest?"

"Leave things as they are. Let the fire burn out."

Legasov shook his head. "The people would not understand. We have to be seen to be doing something."

"But have you considered," asked Fedulenko, "that if you succeed in covering the reactor with sand, it will increase the heat in the core? At the moment it's like a furnace. The air comes in from below and cools the lower layer. It cannot fail to do so. It acts like a grate. But if you block the flow of air, then the heat will increase. This is almost certainly what is happening now. At twenty-nine hundred degrees the fuel will melt, possibly burn through the concrete base, and react with the water in the bubbler pool to produce a far greater explosion than we have seen already."

All the members of the commission now addressed themselves to the horrifying possibility that the molten mass of uranium dioxide fuel, fission products and fuel cladding known as corium, might burn through the concrete base and react with the water below to produce a massive new explosion that would destroy the other three reactors. Even before leaving Moscow, Academician Velikhov had initiated studies at the Kurchatov Institute on the effect of corium on concrete, and he now judged that the corium might already have penetrated halfway through the two-meter slab of concrete.

Even if the bubbler pool could be drained of water, there remained the danger that the candescent corium might burn on down through the foundations and ground to the water table below. This would lead to a further catastrophic thermal explosion, devastating an area of two hundred square kilometers and poisoning the tributaries of the Dnieper River, which provided water for thirty million people in the Ukraine.

The most immediate danger was the water in the bubbler pool. No one knew how much water it contained: the greater the amount, the greater the danger. Someone had to find out, but no one knew how to approach it through the ruins beneath the reactor. The plans of the power station had been lost; the engineers had had to send for those of the similar RBMK at Smolensk. An Armenian engineer volunteered to go on a reconnaissance mission: two soldiers went with him, wearing white protective clothing, respirators and a mask. At 2:00 A.M. on 2 May he returned, haggard, unshaven, wide-eyed, his respirator hanging around his waist. "Less than two hundred cubic meters," he reported, then sank back exhausted into a chair.

Although it was a relief that there was not more, there remained enough water to cause a further thermal explosion. Brukhanov and Fomin, still responsible for implementing the decisions of the government commission, sent the head of the shift and Vadim Grishenka to make their own assessment of the water level. They too saw that the upper bubbler pool was empty, but said that the water that remained in the lower section could only be drained through the gate valves at the bottom.

These had to be opened by hand, which meant someone had to dive into the highly radioactive water. Shasharin, the deputy minister of energy, came up with wet suits and Silayev asked for volunteers, offering lavish inducements as encouragement: a car, a dacha, a flat and a generous pension for the family of anyone who lost his life in this hazardous operation. A detachment of the chemical troops under a Captain Zbrovsky cut a way for the fire hoses, while Yevgeny Ignatenko, the deputy director of the Nuclear Energy Agency, and two engineers put on wet suits, dove into the bubbler pool and opened the valves. The water poured out and the bubbler pool was emptied.

But the danger was not over. The water that had been fruitlessly directed toward the reactor immediately after the accident had flooded the passages and air vents of the basement. Draining this was a more complex operation, undertaken by the commander of a special unit of the fire brigade in Belaya Tserkov, Major Georgi Nagaevsky, who resembled the French actor Jean Gabin, with an honest, square face and deep blue eyes. He was telephoned on the morning of 3 May by his commander in Kiev and asked to find five volunteers for a hazardous operation.

Thirty-five men from Belaya Tserkov had already been to Chernobyl, but since Nagaevsky was the most familiar with the equipment, he himself was the first to volunteer. With four others he proceeded to Ivankov, where he was briefed on the nature and the urgency of the task. He was warned of the danger; they would be working in an area where the dose could be as much as three hundred rems per hour, and they had to avoid any spillage of the radioactive water.

While a normal fire engine could pump water at a rate of forty liters a second, Nagaevsky's two special machines could increase this to 110. They also had hoses up to two kilometers long. Before attempting to lay them in the basement of the reactor, Nagaevsky's men made dummy runs on the uncontaminated ground around Chernobyl, reducing the time it took them to run out the hoses from twenty to seven minutes. They also made precise plans as to how best to approach the fourth unit. There was an entrance designed for the vehicles that delivered the nuclear fuel, and Captain Zbrovsky of the chemical troops cleared this of rubble in one of the armored reconnaissance vehicles.

The operation was postponed; the firemen were unwilling to proceed while the helicopters were still dropping tons of sand and boron onto the reactor, but they continued their rehearsals. By the evening of 6 May, however, they could delay no longer. Before they set out on their mission, the commander of the fire services gave Nagaevsky three liters of vodka, saying, "Give a shot to each of the lads—for courage, better blood circulation and protection against radiation." Nagaevsky declined; however, when he got to the bunker at the power station, a doctor repeated the popular line that it was advisable to drink some alcohol before they set out on their mission, but said that unfortunately they had run out.

Remembering the three liters back at the commission's headquarters in the town of Chernobyl, Nagaevsky asked if anyone would go back to fetch it. A young fireman from Zhitomir volunteered; but when he got back to Chernobyl, his nerve failed him; he said that he had cut his fingers on the plates of the fire engine and so could not return. Another young fireman, Alexander Nemirovsky, was smoking outside in the evening air when the captain on duty came out, shouting, "We need a fireman to drive to the station." Nemirovsky volunteered and drove with a dosimetrist to join the team.

Nagaevsky and his men now drove their fire engines straight under the reactor. Jumping in and out of the protective reconnaissance vehicles to minimize their exposure to radiation, they laid four hoses—two with a diameter of 1,250 millimeters to pump out the water, two with a diameter of 3,000 millimeters to suck out the debris and dirt. Then the contaminated water was pumped out of the flooded basement into a reservoir adjacent to the power station, from which the uncontaminated water had been pumped into the Pripyat River.

The pumps started at 10:00 P.M. The firemen waited in the bunker along with Brukhanov and Fomin. Every fifteen minutes, two men ran out to check the pumps. After half an hour, Nagaevsky reported to Fomin that the water level had decreased. Fomin telephoned Silayev in Chernobyl, whereupon Silayev passed the information on to Ryzhkov in Moscow. However, half an hour later there was no change. Three hours after they had started pumping, the level had decreased by only one centimeter.

At 3:00 A.M., two firemen returned breathlessly to the bunker to report that one of the tracked reconnaissance vehicles of the chemical troops had driven over the hoses, cutting them in twenty different places and smashing the sleeves that connected the sections. Nagaevsky ran out to see radioactive water pouring out from the broken hoses only thirty-five meters from the reactor. The operation came to a halt.

Nagaevsky then rang Chernobyl to tell them to prepare twenty new sections of hose, and Nemirovsky drove back to fetch them as fast as his fire engine would go. New men arrived to help. Dressed in rubber suits and wearing gloves, respirators and masks to protect their eyes, but

standing knee-deep in radioactive water, the firemen replaced the ruptured hoses. Each hose took two minutes, but they finished in under an
hour. They were exhausted, and some already suspected that it was not
just as a result of their exertion. From the start, they had had a strange
taste of sour apples in their mouths, and each had noticed that the others
had dilated, bloodshot eyes. They were afraid; one joked that it would
be worse in Afghanistan, but another said he would prefer that because
at least he would know where the bullets were coming from; here the
danger was intangible and invisible. They only knew that they had been
affected when their voices became squeaky or they started to feel sick.
One of Nagaevsky's men from Belaya Tserkov, started to vomit during
the morning of 7 May. Another man fainted and was sent straight to the
hospital at Ivankov.

By midnight on 7 May, thirty hours after it had started, all the water
had been pumped out of the fourth unit and the immediate threat of a
second thermal explosion had been averted.

7

While the firemen had been working to drain the bubbler pool and
basement, the atmosphere in the commission had worsened, for the
release of radioactive material from the reactor had risen sharply day by
day. The physicists had estimated that only 3.5 percent of the uranium
had been ejected; therefore 96.5 percent remained. The temperature in
the core was estimated to be around 2,300° centigrade. Millions of
curies of radioactive isotopes were still rising into the atmosphere—first
the noble gases, xenon and krypton, then iodine, cesium and some
strontium. Boron was still being dropped by the helicopters, and it was
thought to be absorbing some of the radioactivity, but still the emissions
increased as the Kurchatov scientists had predicted they would—from
two million curies on 1 May to four million curies on the 2nd, five
million curies on the 3rd, seven million curies on the 4th and eight
million curies on 5th.

As the emissions increased, so did the anxiety of the commission.
Calmer than Scherbina, Silayev nevertheless demanded reports every half

hour and passed them on to Moscow. No member of the commission got more than two or three hours of sleep a day. They faced the possibility of not just one but several catastrophes: the pollution of the atmosphere, of the water table, and of the rivers by rain; and further explosions of incalulable force. To avert these disasters, every sinew of the Soviet state was flexed for action. Every branch of the armed forces and its ancillary brigades—the fire department, the civil defense, the troops of the interior ministry, the KGB, the air force, the civilian airline, Aeroflot—were represented at the highest level. Marshals and ministers, chairmen and generals, even the minister of medium machine building himself, the octagenarian Efim Slavsky, all waited on Silayev, straining at the leash to perform acts of heroic valor if only someone would tell them what to do.

But what was to be done? The military engineers under Marshal Oganov had started building a complex system of dams and drains in case rain should wash the radionuclides into the Pripyat. They also drilled boreholes to a level of forty meters where a layer of impermeable clay began, so that if the ground water should become contaminated it could be pumped out before it reached the river; and a start was made on a concrete barrier beneath ground level to prevent water from seeping out from beneath the damaged reactor.

Such long-term measures, however, did nothing to avert the short-term crisis created by the hemorrhage of radionuclides into the atmosphere. No one seemed to know how to prevent it. While Kalugin and Fedulenko, after estimates made with slide rules in the Pioneer camp, insisted that the graphite fire would burn itself out within a period of between a week and a month, the rotund expert on nuclear fusion, Academician Velikhov, declared that there was no obvious remedy, since they faced a problem of an unprecedented kind. Clearly something that was too hot should be cooled down, so he recommended the construction of a heat exchanger beneath the reactor. But this meant digging under the foundations, building a concrete base, and on this constructing a web of pipes through which water could be pumped from the emergency core-cooling system of the other three reactors. It was not a quick solution.

Velikhov's tentative approach to the crisis did not impress the military men like Generals Pikalov and Ivanov. To them he was a bumbling Mr.

Pickwick; Legasov was the man they preferred—lithe, confident, decisive, a leader in the Soviet mold. He backed the idea of the heat exchanger but thought that something more urgent must be done. What about nitrogen? Why not pump it in to freeze the earth beneath the foundations, drive out the oxygen and smother the fire?

For Silayev, it was a matter not of choosing one course of action but of putting all of them simultaneously into effect. The command was given for all available supplies of nitrogen to be brought to Chernobyl. The minister of coal, Mikhail Shchadov, a thickset man with silver hair and a deep, rasping voice, was told to mobilize miners to dig under the foundations of the reactor, and since time was of the essence, metro construction workers, digging the tunnels for the underground railway at Kiev, were summoned to Chernobyl. Every available source of concrete was to be delivered for the construction of the heat exchanger.

8

By this time it had become apparent to the commission that Nikolai Fomin was incapable of acting as chief engineer. Eventually he was replaced by Taras Plochy, who together with Nikolai Steinberg had come from the Balakovsky nuclear power station to help with the emergency.

Working from the bunker beneath the administrative block, they were made responsible for carrying out the commission's instructions to pump nitrogen into the reactor. From the beginning, both thought this an absurd idea. If the explosion had been contained within the structure, it might have made sense, but because the whole unit had been ruptured, all the nitrogen would escape into the open air. However, mere operators could not question the decisions of a man as eminent as Academician Legasov. Another engineer, Sergei Klimov, who had been brought back from retirement in the aftermath of the accident, had already ordered the equipment from Odessa that would convert liquid nitrogen into gas. A special shed had been built to house it behind the administrative block, and at 10:00 A.M. word came from the commission that the machinery would arrive at 2 P.M. that day.

It was delayed. At 4:00 P.M. they were told that it had been brought to the airport at Chernigov by helicopter, and would proceed from there by truck. When it arrived, it would not fit through the door of the shed. They used mallets to smash a wider opening, and by 8:00 P.M. could report to Silayev that they were ready to start pumping as soon as the nitrogen arrived.

This was expected that night but by morning it still had not arrived. They waited all day. At 11:00 that night, Silayev phoned Brukhanov: "Find the nitrogen or you'll be shot." The wretched Brukhanov traced it to Ivankov, where it had been held up because the tanker drivers were afraid to proceed. The problem was quickly solved by a regional commander, who stationed one armored car at the front of the convoy and one at the back, each of them mounted with machine guns. Faced with a choice between visible and invisible threats, the drivers proceeded to Chernobyl. The nitrogen reached the power station at 1:00 A.M.

Steinberg was given the task of checking the supply. He found that two of the twenty tankers were empty, and that five, colored blue rather than yellow, contained oxygen, not nitrogen. Nevertheless, there was enough for the operation to get under way, and by the next evening there was more than enough. Several freight trains had arrived in the station yards at Pripyat; thanks to the power of the commission, most of the nitrogen in the western part of the Soviet Union had been diverted to Chernobyl. Little of it was used, however, because after twenty-four hours even Legasov realized that the operation was a waste of time.

But by then the worst was over. On 6 May, for no apparent reason, the emissions of radionuclides from the reactor suddenly dropped from 8 million to 150,000 curies, and remained at a low level thereafter. By 9 May the fire in the reactor seemed to have gone out. It was Victory Day—the commemoration of the Soviet defeat of the Germans in World War II—and by special government dispensation the restrictions on alcohol that had been ordered to start on 1 May were postponed for the celebration. For the members of the commission at Chernobyl, the struggle to master the reactor had been their own Great Patriotic War, and so it was an appropriate moment to bring out the vodka and celebrate victory in a battle, if not the end of the war.

Still, in the fading light of the evening, a last run by a helicopter over the reactor showed a small but bright spot of red in the crater. Some thought it was just a burning parachute, but Legasov advised Silayev that there might still be a fire in the core. The celebration was postponed. The next day a further eighty tons of lead were dropped into the reactor, after which the luminescence ceased. The vodka was brought out again, and at Ivankov that night victory was celebrated one day late.

VII

1

Working parallel to the Chernobyl commission in Ivankov and the Politburo commission in the Kremlin was the secret medical commission in Moscow. This met daily from 2 May onward at the Ministry of Health to decide on the measures necessary for the care of the victims of the accident and the welfare of the population at large. It was headed by a deputy minister of health, Oleg Shepin, and included Shulzhenko, the head of the secret Third Division, and the health ministers of the Ukrainian and Belorussian republics, but inevitably the most influential members were those who knew most about the effects of radiation. One of these was Academician Leonid Ilyn, who represented the medical commission at Chernobyl; the other was the director of the Department of Hematology at the Central Institute for Advanced Medical Studies in Moscow, Professor Andrei Vorobyov.

Tall, largely bald, alternately arrogant and charming, witty and aloof, Vorobyov was a brilliant scientist who, at this stage in his career, was considered politically sound. But while the career of a man like Ilyn was undoubtedly furthered by his belief in the Soviet system, Vorobyov had good reason to have less forthright ideological loyalties. Both his parents had been scientists, and both had been committed to the Communist cause, but his Russian father, a physiologist, had been shot in 1938 for supposedly plotting the assassination of Stalin, his Jewish mother had spent eighteen years in prison camps, and an uncle, a monk in the Zagorsk Monastery, had spent twenty years in prison. Andrei, aged seven when his father was arrested, was raised first by relatives and then, between the ages of twelve and fourteen, in a state orphanage.

These origins did not impede Vorobyov's scientific career. When invited to join the highly secret Institute of Biophysics in the 1970s, he raised the question of his parents' "crimes" against the state but was reassured that Tupolev and Korolyov, the designers of aircraft and spaceships, had done much of their work in the gulags. Further, Mendel was shunned, Maximov lived in exile, Vavilov was imprisoned, Koltsov was dismissed, Sechenov lived under police surveillance and Korchak had died in a camp.

Working in the Institute of Biophysics, Andrei Vorobyov had helped developed a system of biological dosimetry that could measure a received dose of radiation far more accurately than the traditional method of using a Geiger counter at the scene of the accident. The latter took no account of whether the victim had faced the source of radiation or had only been affected on one side of his body; thus there was no way of knowing which of the organs had been affected.

Biological dosimetry combined an assessment of skin damage with blood and chromosome tests, which enabled doctors to establish the degree of irradiation fairly accurately well before the patient developed the symptoms of acute radiation sickness. Georgi Seredovkin, the doctor who had performed these tests in the hospital in Pripyat on 26 April, had been a graduate student of Professor Vorobyov's, and it was on the basis of his assessment that the first batch of the most seriously affected— twenty-six in all—had been sent to Moscow that same evening on the same plane that had brought General Ivanov to Kiev.

An additional 102 casualties had followed on two further flights the next day. The most seriously affected—the firemen and operators— were sent to the in-patient department of the Institute of Biophysics at Hospital No. 6; others went to Hospital No. 7. When still more patients arrived, irradiated in the aftermath of the accident, more beds were made available in Hospital No. 12. Initially about 300 people required in-patient care, of whom 203 suffered from acute radiation sickness; 129 came to hospitals in Moscow, and the rest were treated in Kiev. By the middle of May, the number hospitalized had risen to 512.

The chief of Hospital No. 6, Angelina Guskova, had forty years experience in treating the victims of nuclear accidents and radiation sickness, but had never before dealt with a catastrophe of this scale. To assist her there was the head of the Hematology Department, a former colleague of Professor Vorobyov's, Alexander Baranov. A tall, bald man with a quiet, gentle manner, he had the melancholy look of someone who had witnessed much suffering and had seen many of his patients die. Beneath him, however, there were only a few junior doctors—mostly young women—who all of a sudden had to lead groups of ancillary medical personnel.

Many of those sent to Moscow by Seredovkin hardly felt ill. Anatoli Dyatlov, Leonid Toptunov, Sasha Yuvchenko, Razim Davletbayev—even Alexander Akimov, whose skin was now a deep brown—had recovered from the faintness and nausea that they had felt in the wake of the accident. In many of them, the worst symptoms at this stage were the swelling and red patches where radioactive water had come in contact with the skin. Guskova and Baranov recognized these as burns caused by beta radiation, which, unlike alpha particles, could penetrate the skin. As the unstable atoms disintegrated, the energy they released in the form of particles and waves had damaged the cells of human tissue and impeded their reproduction. Alpha particles were normally considered less dangerous because they only travel short distances and cannot penetrate the skin, but the doctors knew that if alpha particles were absorbed through wounds or by inhaling radioactive vapor, they could severely damage the lungs and other internal organs.

Finally there were the gamma rays, which could penetrate anything but solid slabs of lead or concrete and so irradiate the whole body. Some radionuclides would be expelled from the body, but others were retained by particular organs—iodine by the thyroid gland, ruthenium by the intestine, strontium and plutonium by the bones, barium and molybdenum by the lower large intestine.

Under the supervision of Dr. Baranov, samples of blood were taken from the patients, cultivated and then studied under a microscope for damage to the chromosomes. From this Baranov could estimate the dose and prescribe the treatment. The greatest danger was to those organs made up of rapidly dividing cells—the skin, hair follicles, the gastrointestinal tract and the bone marrow, which supplies the body with blood. As a result of damage to the chromosomes, the cells, unable to reproduce themselves, would perish. This was particularly lethal when it came to bone marrow. However, death might not be caused simply by exhaustion of the blood; it could come from damage to internal organs—from suffocation, for example, if radiation had destroyed the lungs, or from peritonitis if the intestines disintegrated—and would be accompanied by a number of grievous symptoms: the painful blisters of herpes simplex on the mouth and face; internal hemorrhage leading to bloody diarrhea; high fever; subsequent coma; and an immediate susceptibility to any infection.

Those patients who had received a dose below one hundred rems would be cured by their bodies' own resources. For those who had received doses above six hundred rems, little could be done. It was the patients who had received an intermediary dose for whom the transfusion of undamaged bone marrow from a compatible donor could lead to the regeneration of a healthy supply of blood—or, where a genetically compatible donor was not available, genetically promiscuous bloodproducing cells could be extracted from the livers of human fetuses, which, because of the large number of abortions performed in the Soviet Union, were always in plentiful supply.

This treatment was susceptible to numerous setbacks. The patient's immune system might reject the donor's bone marrow, or the new white blood cells might attack the recipient. Even where a transplant was successful, the patient remained vulnerable to the slightest infection until sufficient new blood had restored the body's immune system. This meant weeks of intensive care in sterile conditions.

There were also great logistical difficulties in preparing bone-marrow transplants for so many patients. The hospital possessed only one blood separator and one centrifuge. The patients' blood types had to be defined before the lymphocyte count dropped too far, and then donors found from among relatives of the patients, blood samples taken and matched. Unfortunately, some of the patients with third-degree radiation disease also suffered from extensive beta burns, so a bone-marrow transplant could not save them; there was nothing the doctors could do but prescribe morphine to ease the pain. On 2 May Dr. Baranov predicted that ten would die; later, as the symptoms of severe internal radiation emerged, he increased this number to thirty-seven.

2

Valeri Khodemchuk had been killed in the initial explosion; Vladimir Shashenok had died in the hospital in Pripyat. The third fatality following the accident at the Chernobyl nuclear power station was not any of the patients who had been flown to Moscow but the good-natured chief of the electrical workshop, Alexander Lelechenko.

Called in to repair the electrical circuits in the turbine hall of the fourth unit, which had meant working knee-deep in radioactive water, Lelechenko had seen that there might be a further explosion of the hydrogen, so with V. I. Lopatuk he had run to close the valves. When, eventually, he had felt dizzy and sick, he had waited for treatment in the hospital in Pripyat, but deciding for himself that his sickness was not serious, had gone back to his flat to rest before returning to work at the station at 4:00 P.M. on 26 April.

At eight that evening, Lelechenko had come home to eat supper with his wife, Lubov, a math teacher, who had never wanted to live in Pripyat because she was afraid of nuclear power. He had told her that he had a dry mouth and a terrible headache and asked for an aspirin. They also drank vodka because he was tense. When she asked him if she should throw away the clothes she had hung out to dry on the balcony, he had told her it was safe to bring them in.

Lelechenko hardly slept that night, tossing and turning on the bed, but the next morning he went back to work at the station. He told his wife that he was needed; "You can't imagine what's going on there. We have to save the station."

"And if there's an evacuation?"

"Go. I'll join you."

Lubov was evacuated to a village near Ivankov, where she continued teaching the children from Pripyat in a borrowed classroom at the local school. When the deputy chief engineer, Lyutov, came to visit his wife, who was billeted in the same house, he told Lubov that her husband was fine. On 30 April Lubov visited her daughter in Kiev, but returned to the village on 1 May. That night, Lyutov woke her to say that Alexander was in the hospital in Kharkov.

Lubov spent much of the following day in the post office, telephoning friends in Kiev and the friends of friends in Kharkov. Lelechenko was not there. Eventually, their daughter Ylena rang from Kiev to say that she had tracked him down at the Institute of Radiology there. "Things are bad," she told her mother. "You must come at once."

Lubov applied for a permit to leave the school and go to Kiev. When she arrived at the Institute of Radiology she was told that her husband was under intensive care. She went up to the special wards. Her husband came out wearing a dressing gown and underpants. His whole body was red. Lubov had brought him an apple, but he could not eat it.

"Cheer up, Sasha," Lubov said. "You'll survive."

"I had to shut off the hydrogen," he said. "If I hadn't shut it off, it would have exploded." He asked her to bring him as much liquid to drink as she could. Then he went back into the ward: he could not stand for long.

Lubov now tried to find a doctor. The first told her that her husband was no worse than any of the others; the second, a woman, confessed that his condition was serious.

"Why wasn't he sent to Moscow?"

"He wouldn't survive the journey."

"What are you going to do for him?"

"We've ordered a blood transfusion."

The next day Lubov Lelechenko went to the office of the Council of Ministers to complain that her husband was not receiving adequate treatment. They assured her that the doctors were doing their best. She returned to the dormitory where her daughter Ylena lived, and together the two women made a special soup. They also brought red wine and caviar, which they took to the hospital with the soup. They were allowed in to feed him; his lips were covered with a yellow oil. A doctor asked Lubov if Alexander had a young male relative who would act as a donor for a bone-marrow transplant.

The next morning—it was now 7 May—Lubov returned to the hospital with more soup. Ylena was waiting in the reception room. The two women went up to the ward. Alexander was not there, and both the chair and the bedside table were gone. A nurse asked Lubov if she was his wife.

"Yes. Is he conscious?"

"The doctor will tell you."

The doctor led Lubov into a small room and told her that Alexander was dead. She wept. They called Ylena to comfort her, then gave her an injection to calm her down. "Please don't cry," the doctor said, "but think about where you would like him to be buried."

3

In Moscow the unprecedented scale of the accident obliged Guskova to call upon resources from outside both the Institute of Biophysics and Hospital No. 6. Not only were the less seriously affected patients sent to Hospitals No. 7 and 12 and the Institute of Radiology in Kiev, but a detachment of Pikalov's chemical troops set up camp in the grounds surrounding Hospital No. 6. The patients themselves were radioactive, and the soldiers washed down the floors of the wards four times a day. There was some danger to the doctors; the more seriously affected patients emitted as much as twenty rems per day from the radioactive cesium in their bodies, and some doctors received doses of as much as two rems per day. It was reported to the government medical commission on 4 May that medical personnel working with irradiated patients had asked for an increase in their pay, and the next day a 25 percent increase, to bring their salaries on a par with those of the radiologists, was approved.

There were offers of help from abroad, but most of them were rejected. On 2 May Professor Vorobyov advised the medical commission against importing stable iodine because its use as a prophylactic was no longer effective. At the same meeting the commission decided to respond to a telegram offering help from the World Health Organization "with a polite 'no.' " Private offers met with the same rejection. As early as 29 April, the celebrated French hematologist Dr. Henri Jammet came to Hospital No. 6 with the scientific attaché from his embassy to offer to fly badly irradiated patients to Paris for bone-marrow transplants. After discussions with Guskova around the polished table in her office, he agreed that the Soviet doctors could cope.

However, on 1 May Guskova received a call from the Politburo to say that an American specialist in bone-marrow transplants, Dr. Robert Gale, would be arriving at the hospital the following day with equipment donated by the American businessman Armand Hammer. Guskova had never heard of Gale. She was familiar with the work of other American specialists from their publications, and if consulted would have preferred their assistance, but the command came from the highest level that this Dr. Gale was to be made welcome.

Even in his own country Dr. Robert Gale was hardly a household name. His great-great-grandfather had been an immigrant from Belorussia called Galinsky. His grandfather had earned his living as a tailor in Brooklyn; it was his father, who worked in insurance, who had changed the family name to Gale. His mother, who came from a similar background, had a brother who had made good in Hollywood and a sister who had married a prominent attorney. His family's saga was similar to that of many other Jewish emigrants from Eastern Europe in the nineteenth and early twentieth centuries.

At the time of the accident at Chernobyl, Gale was an associate professor of medicine at UCLA Medical Center, dividing his time among teaching, administration, laboratory work and patient care, and was also the chairman of the advisory committee of the International Bone Marrow Transplant Registry. The only blot on his record was a censure by the National Institutes of Health for permitting experimental bone-marrow transplants in 1978 without the NIH's approval.

It was not his professional qualifications, however, that brought Gale to the attention of the Soviet authorities in the wake of the Chernobyl accident, but his friendship with the singular American businessman, Armand Hammer. Like Gale's grandfather, Hammer was the child of Jewish immigrants from Russia, and like Gale he had studied to be a doctor. Impelled by the same altruistic impulse that was to inspire Gale, Hammer had traveled to Russia in 1921 to treat the victims of typhus. There he had met Lenin, who, with his New Economic Policy in mind, explained that what his country required was not medical assistance but trade.

Back in the U.S., Hammer arranged for the barter of American grain for Russian minerals and furs, and for the next nine years he worked in the Soviet Union as the representative of American companies. He started a pencil factory in Moscow and an asbestos mine in the Urals. In 1930 he bartered these industrial holdings for Russian works of art, which he sold through his own art gallery in New York.

With a genius for business and trade, Hammer bought holdings in diverse enterprises, among them Occidental Petroleum, which he built into one of the largest oil companies in the world. In 1961, he was back in Moscow as part of an American trade mission, and in the 1970s he secured a large contract for Occidental Petroleum to provide fertilizer to the Soviet Union. As a man whose enterprise had been approved by Lenin himself, the octogenarian Hammer gained unique and privileged access to the highest echelons of the Soviet government and had his own private apartment near the Tretyakov Gallery on Lavrushinsky Street in Moscow.

Hammer was a great philanthropist in several areas, among them cancer research. In September 1984 he had asked the director of UCLA's tissue-typing laboratory, Paul Terasaki, to advise on an application for aid from a scientist at the Hadassah Hospital in Jerusalem. Terasaki had brought in Gale, whom Hammer then invited to fly with him to Israel on his private Boeing, *Oxy One*. It was this trip that forged a friendship between the octogenarian Hammer and the thirty-nine-year-old Gale. Despite the difference in their ages and worldly wealth, they had many things in common. Both were descended from emigrants from Eastern Europe, and it could be said of Gale (as Gale was to write of Hammer), "Some people feel his actions are motivated by flagrant self-promotion."

Gale heard the news of the Chernobyl accident on the radio while he was shaving on the morning of 29 April. He realized at once that after an accident on this scale, there might be a greater demand for bone-marrow transplants than Soviet doctors could supply. He therefore telephoned the director of the International Bone Marrow Transplant Registry, Dr. Mortimer Bortin, in Milwaukee to obtain his approval, as well as that of the other members of the advisory committee, to offer the registry's services.

Knowing how intergovernmental agreements were bedeviled by the posturing of the Cold War, Gale decided to extend the offer through his friend Armand Hammer. Hammer was in Washington for the opening of an exhibition, which he had arranged, of paintings from the Pushkin and Hermitage museums, but Gale reached him by telephone that afternoon, and he agreed to be the conduit. By the end of the day, Gale's proposal had been approved by members of the Senate Foreign Relations Committee. Hammer then wrote a letter to Gorbachev offering the doctor's assistance and the resources of the transplantation registry:

> Dr. Gale is prepared to come immediately to the Soviet Union to meet with Soviet nuclear scientists and hematologists to assess the situation and decide the optimal course of action with the hope of saving the lives of those at risk. I will bear all the costs of his efforts, which can be so important to saving the lives of those citizens who have been exposed.

The letter was delivered to the Soviet Embassy in Washington and also telexed to Anatoli Dubrynin, a member of the Central Committee and a close confidant of Gorbachev's. At 7:00 A.M. on 1 May, Gale received a call from the acting Soviet ambassador in Washington, Oleg Sokolov, extending an invitation to fly to Moscow.

Gale telephoned colleagues around the United States, asking them to be prepared to assemble all the necessary equipment for tissue-typing and bone-marrow transplants. Already a UPI report had been published in the U.S. press that quoted "a Kiev resident with hospital and rescue team contacts," as saying that "eighty people died immediately and some 2,000 died on the way to hospitals." This led Gale to fear that whole families might have been irradiated, making it impossible to find donors from among the near relatives of the victims. Therefore he alerted colleagues as far apart as London and Seattle to register possible donors.

It took Gale only a moment to pack; he traveled often and light. Outside his house, there was already a crowd of reporters: somehow the press had found out what was going on. He caught an afternoon flight to Europe from Los Angeles airport and reached Moscow at 6:00 P.M. the next day.

4

Obedient to the political imperatives that for a good Communist took precedence over the demands of his profession, Dr. Baranov was diverted from the care of his patients to exchange his white hospital overalls for a shirt, tie and dark suit and proceed to the Sovietskaya Hotel to welcome Dr. Gale. Protocol was protocol. He waited in the grandiose lobby, whose thick red carpets, antique furniture, oil paintings, chandeliers and marble staircase were a marked contrast to the utilitarian surroundings of Hospital No. 6, until Gale arrived from the airport. Gale was flanked by two "minders," Victor Voskresenski, a former medical attaché in the Soviet embassy in Washington, who acted as his interpreter, and Nikolai Fetisov, an official from the Ministry of Health. If Baranov was taken aback by the sight of the trim American wearing wooden clogs he did not show it. The four men went up to the suite that had been reserved for Gale, and over supper Baranov briefed him on the patients in his care.

At nine the next morning, after breakfasting with him at the hotel, Baranov brought Gale to the hospital. The American had risen at six and jogged eight miles; the chain-smoking Baranov took him up to his office on the sixth floor, where he was introduced to Angelina Guskova.

Guskova was disappointed. Vorobyov had led her to believe that Gale would arrive with much-needed equipment provided by Hammer, but she saw only an American doctor carrying a small bag. Nevertheless, she welcomed him as protocol and party discipline required, before, accompanied by Voskresenski, the interpreter, she left Baranov to take Gale on a tour of the wards. Later that morning the American attended a conference chaired by Guskova on the condition of the various patients. She had a brisk, abrupt manner and was impatient with any dissension among her subordinates; the need to translate what was said into English irritated her because it protracted the proceedings, but she was satisfied by Gale's modest demeanor.

After lunch at the Sovietskaya with his two minders, Gale returned to the hospital, where Baranov invited him to assist in the extraction of bone marrow from a donor. Before Gale's arrival Baranov had already performed four transplants, and another was scheduled for later that day, but after the one operation he insisted that Gale had done enough and brought him back to his office on the sixth floor.

There Gale drew up a number of recommendations, which were then presented to Guskova and by Guskova to the medical commission. They should invite two of Gale's colleagues from California—Paul Terasaki (who would bring a mobile tissue-typing laboratory to Moscow) and Dick Champlin, a clinical specialist—and a third specialist from Israel, Yair Reisner. Gale also gave Guskova a list of equipment that should be sent from the United States.

That evening, Baranov, Gale and his entourage dined with the cardiologist Yevgeni Chazov, chairman of the Soviet committee of Physicians for the Prevention of Nuclear War. After caviar, sturgeon and innumerable toasts in brandy and vodka to peace, friendship and the smallness of the world, Chazov took Gale to the Orthodox Easter Vigil—a glimpse of Russian folk culture for a distinguished guest.

Back at his hotel at 2:00 A.M., Gale unwound by listening to Beethoven's music on his portable hi-fi and musing over the momentous events of the day.

Gale's recommendations were accepted, and in the days that followed Terasaki, Champlin, and finally Reisner arrived—uneasy, because he was an Israeli citizen and there were no diplomatic relations between Israel and the Soviet Union. Although none had any experience in the treatment of radiation sickness, they had developed the most sophisticated techniques for bone-marrow transplants and could both operate and obtain the most advanced drugs and equipment available on the world markets, which, because of their cost in hard currency, had hitherto been beyond Baranov and Guskova's wildest dreams. Now, after a call from Gale to Hammer's assistant in Los Angeles, state-of-the-art equipment was being flown to Moscow.

The equipment was of more value to the Soviet doctors than the skills of the foreign physicians. On 4 May the French ambassador offered the services of Professor Georges Mathé. The deputy minister of health, Shepin, alarmed by the arrival of Gale's colleagues, insisted that no more foreigners should be invited without the commission's approval. Vorobyov told the medical commission that Gale was a better hematologist than Mathé, so the French government's offer was turned down.

At Hospital No. 6, an amicable working arrangement was quickly established between the Soviet doctors and Gale's team. Guskova was impressed by Gale's ability to get things done, and she liked Terasaki's quiet, modest manner, as well as his Japanese equipment. His work was invaluable in testing the blood of the hundred potential donors required for thirteen transplants and in checking the analyses already made by their own labs. Baranov got on well with Champlin; they vied with one another about who could best remember various articles published in specialist journals. Both Baranov and Guskova were encouraged by the arrival of foreign drugs, equipment and expertise to perform bone-marrow transplants on patients, even where their diagnosis suggested that they were likely to die anyway as a result of extensive burns and internal radiation.

5

There was a tragic correlation between the heroism of the injured and the seriousness of their disease. The two young lieutenants of the fire brigade, Pravik and Kibenok, were horribly irradiated. Before the arrival of his commander, Major Teliatnikov, Pravik had directed his men in extinguishing the many fires that broke out in the wake of the accident. Wearing nothing but the standard-issue uniform of a Soviet fireman, he and his men had spent several hours exposed to radioactive eruption— first on the roof of the turbine hall, then in the reactor hall and finally when turning their hoses on the burning reactor itself. With their boots sticking to the melting tar on the roofs, breathing in smoke from the burning graphite, they were neither warned nor sought to discover what risks they ran in fighting the fire. They had done their duty until they had succumbed to faintness and nausea and were taken to the hospital in Pripyat.

Of the 129 patients admitted to Hospital No. 6, 111 developed acute radiation sickness. The most severely afflicted, such as Pravik and Kibenok, were placed in individual wards on the eighth floor, which were subject to a rigorous antiseptic regime; there was ultraviolet sterilization of the air and strict rules for the nursing and medical staff, including washing their hands before entering the ward, wearing individual overalls and masks, and treating their shoes in a mat soaked in an antiseptic solution. As the disease progressed, the firemen's bodies became dark and swollen, and the painful blisters of herpes simplex appeared on their faces, lips and the inside of their mouths. They could not eat or drink, their temperatures rose and they suffered atrocious pain before passing into a coma.

Many of the families of the victims who had followed them to Moscow stayed in a hostel built for the relatives of patients opposite the hospital. The facilities were basic, and because there were so many family members visiting, they slept four or five to a room. As a rule, they were not allowed into the wards, but would congregate every morning in the single-story building at the entrance to the hospital precincts. Many of the wives were distraught. Their husbands were suffering; some would certainly die. They had also lost their homes and belongings, and their children were scattered, staying either with grandparents or in holiday camps. Most of their clothes had been left in Pripyat, and they had difficulty in getting hold of any money; they either had to borrow from friends or go begging to the Ministry of Energy.

Some had ailments of their own. The woman who had gone to work at the power station after the accident suffering from an unusually heavy period and had inadvertently dropped a length of cotton wool in the radioactive dust on the corridor floor and then used the cotton as a sanitary napkin now found that the skin of her groin was peeling away and her pubic hair was falling out.

The lobby at the entrance to the hospital became a concourse for the exchange of sympathy and information. Here families would be given the most recent bulletins on the health of their relatives or would discuss their condition with the doctors. Unofficial intelligence was supplied to the wives by Anatoli Sitnikov's wife, Elvira, who had managed to get a job in the hospital. She would take messages and newspapers to the patients and would later report on their condition to their wives.

During the first few days in Moscow, some of the patients had felt well enough to get out of their beds and talk to their friends, or to uninvited visitors who found their way into the wards. Sasha Yuvchenko awoke one morning to find an officer of the KGB waiting to interrogate him.

Since smoking was not permitted in the wards, many of the patients went out onto the landing. There Dyatlov met up with Sitnikov, Chugunov and Orlov, three of his old friends from Komsomolsk, and with some of the younger engineers, such as Alexander Akimov and Razim Davletbayev, who had been with him in the control room of the fourth unit. They discussed the accident, but no one raised the question of who was to blame. They could not understand what had happened. "We are open to any suggestion, lads," Dyatlov said to the young operators. "Don't be afraid to come out with even the most farfetched ideas."

Young Leonid Toptunov, who had been at the controls at the time, had his father constantly by his side, but since he had only followed instructions, he had less reason to be tormented by misgivings than either Dyatlov or Alexander Akimov, the head of the shift. However, the older and more experienced Dyatlov not only had received a lesser dose of radiation, but also had better built-in protection against scruples and self-doubt than the unfortunate Akimov, who, though he kept repeating that he did not know what had gone wrong, nevertheless felt responsible and confessed to the shift foreman, Victor Smagin, that his conscience hurt him more than the pain.

Akimov's condition quickly deteriorated, but even as it worsened he was tormented by not knowing what had gone wrong. "How could it have happened?" he asked his friend Razim Davletbayev, who visited him in his sterile room. "What did we do wrong? Everything went well until we pushed the AZ button." Akimov knew he might die. "My chances are slim," he said to Razim, and to prove it he pulled out a tuft of his hair. "But if I do survive, one thing is for sure: I'll never go back to work in the nuclear field. I'll do anything . . . I'll start my life from scratch, but I'll never go back to reactors."

Although no visitors were supposed to enter the antiseptic zone, individual doctors made exceptions for those who were likely to die. Akimov's mother brought him a thermos of chicken broth, but when she saw him she fainted. The first time his wife, Luba, came to visit him she found him methodically pulling out the hair from his head and throwing it in a bin. She tried to raise his spirits by describing how, when he got better, they would live by a river and earn their living, like Dyatlov's father, by regulating navigation and checking buoys. She had brought drawings by their children to distract him, and she told him how well they were doing in school. Once she looked back from the window and saw that her husband was now pulling out tufts of his moustache.

"Don't worry," he said to her, "it doesn't hurt."

To witness so much suffering in those they loved took a terrible toll. Luba, aged thirty-three, suddenly looked old; once a nurse took her for Akimov's mother. The only consolation came from the companionship of the other wives. Luba was an old friend of Inze Davletbayev's; she had invited Inze to stay in her dormitory in Moscow when she had come from Bawly in Tatary to marry Razim, and when Inze had fallen ill, she had nursed her. Now, every morning, the two women heard from Elvira Sitnikov about their husbands' worsening condition. A doctor had told Luba that radiation harmed the reproductive organs. Luba thought that they might be unable to have more children, and she told Inze how one evening she had seen her husband naked, his skin now dark, his penis rotting and black.

Akimov had two younger twin brothers; one of them gave bone marrow for a transplant, but new bone marrow could not arrest the disintegration of his flesh. He had atrocious beta burns and because radioactive vapor had rotted his lungs, he had great difficulty in breathing. His bowels and intestines disintegrated too; excrement oozed out of his body as a bloody diarrhea. Other operators suffered from the same dreadful condition. Brazhik's body was also rotting. A tall, well-built young man with curly hair, he had been a keen sportsman back in Pripyat. Too shy to ask a nurse for a bedpan, he had gotten out of bed one morning and found that the skin of one leg slipped down to his ankle like a loose sock.

Hooked up to drips giving intravenous antibiotics, the patients required constant nursing, the burns and blisters on their skin needing antibacterial, fungicidal and antiviral therapy. The severely irradiated patients were themselves radioactive, and the nurses in the wards more than earned their extra pay. As the disease progressed, the victims' temperature rose—whether from infection or radiation it was hard to tell—and finally they lapsed into a coma.

The first to die, on 10 May, were Kibenok, Pravik and another fireman. Akimov followed soon after, Luba at his side. Anatoli Kurguz, an operator scalded by radioactive steam, died on 11 May; on 12 May, Alexander Orlov, Chugunov's deputy, and one more fireman died; three more men followed on 13 May, including Alexander Kudriatsev, one of the two young operators who had been sent by Dyatlov to lower the control rods. On 14 May twenty-six-year-old Leonid Toptunov died. He had been at the controls of the reactor and afterward had followed Akimov in his fruitless quest to get water into the reactor. Another fireman followed the next day.

Few of those who were doomed knew that they were going to die, and Guskova did not tell them. Arkadi Uskov learned that Akimov was dead from one of the soldiers cleaning his ward. Razim Davletbayev read about it in the papers some days later. Inze Davletbayev, who had had to leave Moscow for Bawly because one of her sons was ill, returned to the hostel to find that Luba Akimov had left. She telephoned Elvira Sitnikov, who told her that Alexander had died.

As a result Inze became distraught; she was sure that Razim, too, was going to die and felt well up inside her a longing for spiritual solace. She took a tram to the center of Moscow, then walked to the Tretyakov Gallery, hoping to find some consolation in its magnificent works of art. The gallery was closed. She walked back toward the center of the city and found herself under the walls of the Kremlin. Without considering what she was doing, she joined the line to visit Lenin's tomb. She waited for two hours, tears trickling down her lovely Tatar face. Finally she came to the mausoleum and went past the erect guards and down the black marble stairs to the burial chamber of polished red-flecked granite. There, in a glass coffin, lay the mummified body of Lenin, his eyes closed, his features in repose. Four soldiers with fixed bayonets stood at the corners of the marble bier. Inze shuffled past, staring at the waxen face of her country's legendary leader, and then, without premeditation, found herself praying to Lenin that all would be well.

6

Baranov's original estimate of ten deaths had been proved wrong, and even the revised figure of thirty began to seem optimistic. Many remained seriously ill—not only personnel from Chernobyl like Razim Davletbayev, Vladimir Chugunov, Anatoli Sitnikov, Piotr Palamarchuk and Victor Proskuriakov, but also accidental bystanders like Georgi Popov, a turbine engineer from Kharkov who had waited for an hour outside the power station for a bus to take him to Pripyat, and a solitary middle-aged woman, Klavdia Luzganova, a security guard who had remained at her post.

None of the deaths was due directly to the destruction of bone marrow: 70 percent came either from extensive burns—both thermal from escaping steam and beta from radioactive water—or from the disintegration of the lungs and intestines. Had the first victims survived these injuries, the bone-marrow transplants might have saved their lives, but there were only nineteen patients whose doses warranted the complex operation, and by 12 May the last had been performed.

This meant that as far as Guskova and Baranov were concerned, the work of Gale and his team was done, and as an expression of gratitude, the team was offered a two-day excursion to Leningrad. By now Gale's wife, Tamar, had joined him from Los Angeles, and with Dick Champlin and Yair Reisner she left for Leningrad on the night train. But Paul Terasaki had to leave Russia to deliver a lecture in Paris, and Gale himself declined the offer for the high-minded reason that he did not want to abandon the victims of the world's worst nuclear disaster to go sightseeing in Leningrad.

He also had a new role in mind. Before leaving the United States, Armand Hammer had impressed two things upon him: first, that he must convince his Soviet colleagues that he was only there to help them; second, that he must not speak to the press. When he had arrived in Moscow, his hosts had also asked that he not brief reporters, and when he had been called by journalists at his hotel he had kept his word. But to Gale, this reticence was emphatically un-American. A disaster of this magnitude in the United States would have brought out an army of reporters, photographers and television cameramen. There would have been regular briefings from the hospital, with Gale himself probably appearing on all the networks.

The Soviet government had released some information. On 6 May there had been a press conference at the Ministry of Foreign Affairs where questions were answered by, among others, Boris Scherbina, the chairman of the government commission: A. M. Petrosyants, chairman of the State Committee for the Use of Atomic Energy; Dollezhal's deputy from NIKYET, Ivan Yemelyanov; and Yuri Israel's deputy from the State Committee of Hydrometerology, Y. S. Sedunov.

Questions on casualties and the condition of the victims were handled not by Andrei Vorobyov but by his namesake, Yevgeny Vorobyov, deputy minister of health. Before he spoke, Scherbina had already mentioned "with satisfaction that Professor Gale and Professor Terasaki, who had arrived from the United States, are helping Moscow specialists." Later in the conference, Yevgeny Vorobyov reported:

> . . . only two people were killed. One of these died from heat burns. The burns covered eighty percent of his body, but in spite of all the measures that were taken, he died. These burns were absolutely huge. The other person died from injuries incurred from things falling on him. Those are the two people who died in the first twenty-four hours of the accident. A mere two hundred and four people were taken to hospital after being diagnosed with radiation sickness—two hundred and four people, that is, with varying degrees of contamination from radiation. . . . Of these two hundred and four people, eighteen people were diagnosed as having a severe degree of contamination.

Vorobyov also announced:

> . . . with satisfaction that Professor Gale from the United States has arrived in the Soviet Union and is giving consultative help. He is a great expert in the field of bone-marrow transplantation. We hope that Professor Gale, apart from his own experience, will be able to give other help to the victims, including a number of medicines.

For Gale, such reticent announcements were not enough to dispel the rumors of catastrophic casualties that were rife in the United States, and he felt that he, Robert Gale, was in a unique position to set the record straight. While his wife was away in Leningrad with Champlin and Reisner, he raised the matter with his minder from the Ministry of Health, Victor Voskresenski. He conceded that he had agreed not to talk to the press, but he felt that by keeping silent indefinitely he and his team

Igor Kurchatov. COURTESY OF THE KURCHATOV INSTITUTE

Academician Anatoli Alexandrov. COURTESY OF THE KURCHATOV INSTITUTE

Academician Nikolai Dollezhal. COURTESY OF NIKOLAI DOLLEZHAL

Academician Valeri Legasov.
COURTESY OF VALENTIN OBODZINSKI

The turbine hall.
COURTESY
OF NIKOLAI STEINBERG

Vadim Grishenka
at the controls.
COURTESY
OF VADIM GRISHENKA

The first mobile homes at Pripyat.
COURTESY OF NIKOLAI STEINBERG

Alexander Akimov (*left*)
with Nikolai Steinberg (*right*).
COURTESY OF NIKOLAI STEINBERG

Party leaders in Pripyat (*left to right*): Vasili Kizima, head of construction;
A. S. Gamanyuk, Pripyat party secretary; Vladimir Voloshko, chairman
of the town council. COURTESY OF NIKOLAI STEINBERG

Nikolai Steinberg with
Katya Litovsky.
COURTESY OF NIKOLAI STEINBERG

Aerial view of the
V. I. Lenin Nuclear Power
Station at Chernobyl.
The ruined fourth reactor
is in the foreground;
the administrative block
is in the background.
COURTESY OF NIKOLAI STEINBERG

Dr. Angelina Guskova.
PHOTO BY PIERS PAUL READ

Construction of the protective wall.
COURTESY OF DMITRI SHATALOV

Dr. Alexander Baranov.
PHOTO BY PIERS PAUL READ

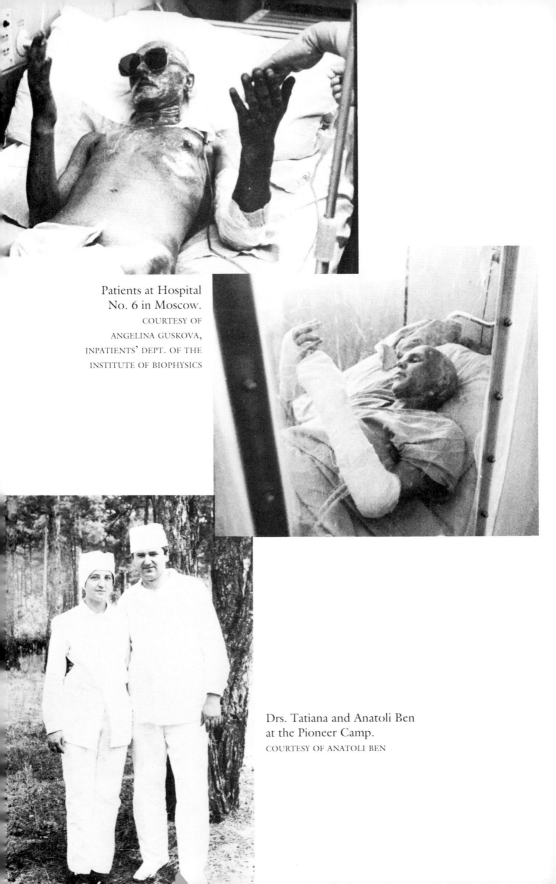

Patients at Hospital
No. 6 in Moscow.
COURTESY OF
ANGELINA GUSKOVA,
INPATIENTS' DEPT. OF THE
INSTITUTE OF BIOPHYSICS

Drs. Tatiana and Anatoli Ben
at the Pioneer Camp.
COURTESY OF ANATOLI BEN

General Pikalov
at Chernobyl.
COURTESY OF
GENERAL PIKALOV

General Pikalov
eats the food served
to his men.
COURTESY OF
GENERAL PIKALOV

Military vehicles
abandoned in the
thirty-kilometer zone.

The Ka-27 helicopter and crew.
COURTESY OF EDWARD KOROTKOV

Academician Leonid Ilyn.
COURTESY OF LEONID ILYN

The sarcophagus completed.
COURTESY OF NIKOLAI STEINBERG

Lubov Kovalevskaya as a child (*second from left*). The stockade of the prison camp can be seen in the background.
COURTESY OF LUBOV KOVALEVSKAYA

Lubov Kovalevskaya in Pripyat.
COURTESY OF LUBOV KOVALEVSKAYA

Pripyat, May Day, 1985. Vladimir Chugunov and Anatoli Dyatlov (*first and second from left*) take part in the parade.
COURTESY OF VLADIMIR CHUGUNOV

Efim Slavsky in retirement. PHOTO BY PIERS PAUL READ

Grodzinski's
mutations.
COURTESY OF
DMITRI GRODZINSKI

Professor Yuri Israel.
PHOTO BY PIERS PAUL READ

Dr. Anatoli Romanenko.
PHOTO BY PIERS PAUL READ

Judge Raimond Brize.
COURTESY OF RAIMOND BRIZE

Public Prosecutor
Yuri Shadrin.
COURTESY OF
YURI SHADRIN

Victor Brukhanov,
Anatoli Dyatlov,
and Nikolai Fomin
in the dock.
COURTESY OF
RAIMOND BRIZE

Anatoli Dyatlov
on the witness stand.
COURTESY OF
RAIMOND BRIZE

Irina Avramenko, resident
of the thirty-kilometer zone.
PHOTO BY PIERS PAUL READ

The fence around
the thirty-kilometer zone.
PHOTO BY PIERS PAUL READ

Anatoli Dyatlov after
his release from prison.
PHOTO BY PIERS PAUL READ

Victor Brukhanov after
his release from prison,
with his wife, Valentina.
PHOTO BY PIERS PAUL READ

Vladimir Gubarev.
PHOTO BY PIERS PAUL READ

Professor Andrei Vorobyov.
PHOTO BY PIERS PAUL READ

Pripyat today. COURTESY OF NIKOLAI STEINBERG

Left to right: Edward Korotkov, Nikolai Melnik and Igor Erlich
in the cabin of the Ka-27 helicopter.
COURTESY OF EDWARD KOROTKOV

were losing all credibility. Surely the time had come for a conference of some kind? If would be the best way to reassure the outside world.

Voskresenski passed on Gale's suggestion to his superiors on 12 May. Since no one thought that the American could be relied upon to exercise the same discretion as a Soviet doctor, the question had been raised the day before at the meeting of the medical commission, which was chaired by the deputy minister of health, Shepin. The commission already faced a plethora of problems—the setting of intervention levels, the decon- tamination of radioactive milk and foodstuffs, further evacuation from an extended zone. In addition, there were dangerous symptoms of growing panic caused by the setting up of dosimetric checkpoints at railroad stations and airports, and there was a scarcity of trained specialists to reassure the population.

In such a crisis it appeared imperative to control the flow of informa- tion. The commission discussed the question of how much and what kind of data should be disclosed. It was agreed that A. M. Petrosyants, from the atomic energy committee, and Andrei Vorobyov, the hematol- ogist who sat on the commission, should act as spokesmen. They should admit the deaths that had taken place in Hospital No. 6, because the American specialists already knew about them, but any further informa- tion the Soviet press possessed must not be published without first being cleared by the Ministry of Health.

On the morning of 14 May, Gale attended a meeting with Ivan Nikitin, chief of protocol at the Ministry of Health, together with Nikolai Fetisov, an official from the ministry, Voskresenski and Professor Vorobyov. He was told that he and Vorobyov could answer written questions, but the idea that Champlin and Reisner should appear on stage was turned down. That afternoon Vorobyov reported to the medical commission, which had been told the day before that Petrosy- ants had refused to take part; therefore it had been left in the hands of Professor Vorobyov. The deputy minister, Shepin, repeated that only information about patients in Hospital No. 6 was to be released, since Gale knew it in any case. Vorobyov reported that Armand Hammer wanted to visit the hospital, Shepin said that this was undesirable, but that the question should be settled by those who had invited Hammer in the first place.

7

That evening, 14 May, nearly three weeks after the event, Mikhail Gorbachev spoke to the nation for the first time about "a misfortune that has befallen us: the accident at the Chernobyl nuclear power plant." In describing the accident and its aftermath, he occasionally strayed into areas of wishful thinking and when it came to casualties he was economical with the truth. He told his audience that "The inhabitants of the settlement near the station were evacuated within a matter of hours," and that nine men had already died.

> Two died at the time of the accident—Vladimir Nikolayevich Sha-shenok, an adjuster of automatic systems, and Valeri Ivanovich Khodemchuk, an operator of the nuclear power plant. As of today two hundred and ninety-nine people are in the hospital diagnosed as having radiation disease of varying degrees of gravity. Seven of them have died. Every possible treatment is being given to the rest. The best scientific and medical specialists of the country, specialized clinics in Moscow and other cities are taking part in treating them and have at their disposal the most modern means of medicine.

Gorbachev also expressed

> our kind feelings to foreign scientists and specialists who showed readiness to come up with assistance in overcoming the consequences of the accident. I would like to note the participation of the American medics Robert Gale and Paul Terasaki in the treatment of the affected persons, and to express gratitude to the business circles of those countries which promptly reacted to our requests for the purchase of certain types of equipment, materials and medicines.

Having thanked Gale and the director general of the International Atomic Energy Agency, Hans Blix, Gorbachev turned in fury on the media and the governments of the West, which had used Chernobyl as a pretext to launch "an unrestrained anti-Soviet campaign":

It is difficult to imagine what was said and written these days—
"thousands of casualties," "mass graves of the dead," "desolate
Kiev," that "the entire land of the Ukraine has been poisoned,"
and so on and so forth.

All this black propaganda Gorbachev saw as a pretext

to defame the Soviet Union, its foreign policy, to lessen the impact
of Soviet proposals on the termination of nuclear tests and the
elimination of nuclear weapons, and at the same time to dampen
the growing criticism of the U.S. conduct on the international
scene and its militaristic course.

He suggested that the accident at Chernobyl and the reaction to it were
a test of political morality that "the ruling circles of the U.S.A. and their
most zealous allies"—particularly the West Germans—had failed. How
could they suggest a Soviet cover-up when

Everybody remembers that it took the U.S. authorities ten days to
inform their own Congress and months to inform the world community about the tragedy that took place at Three Mile Island
atomic power station in 1979.

Gale therefore prepared for his press conference the next day in an
atmosphere of ideological animosity. The first obstacle was Armand
Hammer, who wanted to visit Hospital No. 6. At first the Soviets
refused. Hammer insisted, the Soviets relented, and the chairman of the
medical commission, Oleg Shepin, was there to receive him.

Hammer, Gale and their entourage of officials then drove to the
Ministry of Foreign Affairs. It had been decided at the last minute that
Hammer also should appear on the platform with Gale and Andrei
Vorobyov. Sitting center stage, Gale commended the work of his Soviet
colleagues, Professor Vorobyov, Professor Guskova and Dr. Baranov.
He gave the number of patients treated and transplants performed, while
Vorobyov described the medical monitoring service in the area around
Chernobyl. Vorobyov named a dose of 100 rads as that "beyond which
there is a health danger. Only persons in the immediate vicinity of the

Chernobyl reactor at the time of the explosion received radiation exposure in excess of one hundred rads." Vorobyov also reassured the journalists that they were holding nothing back. "We report everything we know in this field to the IAEA," he said, "and everything we report can be easily checked."

There was applause when Hammer said that the equipment that had been flown to Moscow was his gift to the Soviets, and some interest when Gale announced, in answer to a question, an item of news that he himself had just received: that he and Hammer were to meet Gorbachev himself later that day.

From a Soviet perspective, the conference went well. Given their constant and continuing struggle to persuade public opinion in the West of the dangers of nuclear war, it was valuable to hear from Gale that "people who believe meaningful medical assistance is possible for the victims of a nuclear war are mistaken."

The invitation to visit Gorbachev had been sprung on Gale in a note handed to him during the press conference by Nikolai Fetisov. He was not averse to the role he now seemed cast in as a pawn in a diplomatic game of chess. His friendship with Hammer had introduced him to the trappings of wealth and power. When he had first traveled in *Oxy One,* Hammer's private Boeing 727, he had observed the three separate bathrooms with gold taps and marble-topped sinks, the well-stocked refrigerator and microwave oven in the galley, and the comfortable sofas and armchairs in the different lounge areas, as well as a dining-room table that could seat eight. So, too, in Moscow, he had noticed Hammer's beautiful paintings and furniture, as well as the Western television and video recorder, all in stark contrast to the penurious simplicity of even the most senior Soviet citizen's living quarters.

Yet Gale was neither spoiled nor intimidated by his sudden promotion, and even when giving a press conference in the Soviet Foreign Ministry, or as he walked down the ultimate corridor of power in the Kremlin, he wore his wooden clogs over bare feet. Well might the two army officers who escorted him look askance at such flagrant expressions of what a man could get away with in the land of the free.

A number of photographers stepped forward to take pictures as Hammer and Gale were greeted by the man Gale considered the most powerful person in the world. Wondering with a physician's professional interest why a man with such influence did not have a simple skin graft to remove the birthmark on his forehead, Gale took a seat at a long conference table and listened to Gorbachev thank him and Hammer for their help. Whatever the original motive for accepting Hammer's offer, it was politic now to contrast the humanitarian help of individual Americans with their government's dastardly exploitation of the tragedy for Cold War propaganda.

Hammer did his best to explain that in the United States the government has no control over the press. Gorbachev was not satisfied. Secretary of State Shultz had accused the Soviets of lying about the casualties; Gorbachev felt President Reagan himself wanted to "poison the hearts of the world against the Soviet people, and divert attention from warlike American actions in the Persian Gulf and the Gulf of Sidra."

It seemed to Gale that Gorbachev was genuinely angry. Gale tried to placate him by saying that his presence was proof that the American people took no pleasure in the suffering of innocent people. "But what if Chernobyl had been in space?" Gorbachev went on. Was it not a warning of what could happen if the Americans continued with the development of their Strategic Defense Initiative?

When Hammer suggested a meeting with Reagan, Gorbachev countered that there was no point in meeting if there was nothing to discuss. The hour allotted to the meeting drew to an end. As they rose to leave, Hammer raised the question of the plight of the Jews in the Soviet Union. "These people don't want to be in your country. They're giving you a hard time and a lot of bad press. Why don't you let the Jewish people go?"

"Dr. Hammer," Gorbachev replied. "Jewish citizens like it in the Soviet Union. In many countries, they are discriminated against, but here there is no anti-Semitism. Our people are happy. Whatever else you hear is Western propaganda."

That evening on Soviet television the news program "Vremya" gave wide coverage to the meeting between Gale, Hammer and Gorbachev and was followed by a broadcast of the entire press conference that had been held that afternoon. All at once Gale was a celebrity: he returned to his hotel musing that never before had a private American citizen ever received as much exposure in the Soviet Union as he had that night.

The next morning Gale went to Hospital No. 6 to take his leave of Guskova and Baranov. They had watched the press conference on television the night before; so too had the patients, and hearing Gale say "We anticipate more deaths" had done nothing to improve their morale. For the doctors and nurses, it was work as usual: both Guskova and Baranov wore their white physicians' coats. Gale told Guskova how wonderful it had been working with them and how much they had appreciated his help. They shook hands and Baranov escorted him down to his car. "Thanks a lot," he said in his halting English and with one of his habitually mournful smiles.

Later that day, Gale flew off to California with Tamar, Gail Reisner and Armand Hammer in the marble-topped, gold-plated luxury of *Oxy One*.

8

Guskova and Baranov had less reason to feel elated than Robert Gale. Among the patients under their care, the dying continued. That very day, the young Victor Proskuriakov, who had been sent by Dyatlov to lower the control rods, died from his skin injuries and post-transfusion shock. So too did Lopatuk, who had so courageously helped Lelechenko repair the electrical circuits. There followed a lull of only a single day before another engineer succumbed to his thermal and radiation burns. On 18 and 19 May, three more men followed.

Besides the suffering of the patients, Guskova, Baranov and the team of doctors were faced with the anguish of their relatives. Their presence at the bedside of the dying was left to the doctor's discretion. One operator's pregnant young wife asked Guskova if she could see her husband. He was suffering terribly from the effects of radiation, with heavy hemorrhaging in the thorax and the abdomen. Guskova was afraid of how she would react to such a terrible sight and told her to come back the next day.

"Will you give him this magazine?"

Guskova took the magazine. When the girl left, the doctor returned to her patient. Suddenly he opened his eyes. "Was that my wife?"

"Yes."

"What did you tell her?"

"To come back tomorrow."

"Good." A few moments later he died.

Beyond the emotional stress of witnessing so much suffering and sorrow, Guskova and Baranov felt considerable professional disappointment as their patients died one after another. With the help of Gale, Champlin and Reisner they had performed thirteen bone-marrow transplants, six with embryo liver cells. Some had been done despite the original prognosis that death was inevitable after a dose of six hundred rems; Akimov's dose, for example, was over fifteen hundred rems. It was thought better to do something than nothing, particularly after the arrival of Gale and his team with their sophisticated equipment; yet the whole process caused stress to the donors, and in six of the cases, where there were severe beta burns, it turned out to have been a wasted effort.

Some who had not been badly burned died from the damage done by radiation to their internal organs. Between 20 and 25 May, two patients died from suffocation and one—a female security guard—from heart failure. On the 29th, more than a month after the accident, Anatoli Sitnikov succumbed to a posttransplantation immune depression that left him unable to resist an infection.

Sitnikov's death was particularly disheartening to all those from the Chernobyl nuclear power station. His wife, Elvira, had been the mainstay of the younger wives, and Sitnikov himself was much liked and admired. When sent by Fomin to inspect the damaged reactor, he had looked straight into the crater from the roof of the water-treatment plant and had received a massive dose of gamma radiation. Vladimir Chugunov, Sitnikov's friend from Komsomolsk, lying seriously ill in a different ward, was convinced that Sitnikov had died as a result of the bone-marrow transplant. Chugunov had bad beta burns on his legs, had lost all his hair, and his weight had fallen to 88 pounds, but he was nevertheless convinced that he should resist the disease with his body's own resources. The doctors complained to his wife that he was an intolerable patient, refusing to take the pills they prescribed and being fussy about his food.

Sasha Yuvchenko, who in the wake of the accident had first helped his superior, Valeri Perevozchenko, to open the valves in the fruitless attempt to get water to the reactor, had bad beta burns on his arms where his skin had come into contact with the radioactive water. Fortunately he had built up considerable muscle from rowing on the Pripyat, so in performing skin grafts the doctors were able to cut out tissue 3 to 4 centimeters thick and thus save his arms from amputation. Razim Davletbayev's condition grew worse; his temperature rose, he felt weak and he was constantly shivering with cold. Sores appeared on the inside of his mouth; blood seeped out of his nose. His hair turned gray but did not fall out.

The huge Piotr Palamarchuk, who had carried the injured Shashenok from the ruins of the fourth unit, had atrocious beta burns on his shoulders where they had touched the body of the wounded man, soaked in radioactive water. The prognosis for Piotr was poor; besides the beta burns, he had received a dose of 780 rems. His wife, Tatiana, was asked to find a donor from within the family; although terrified, his sister volunteered and came to Moscow from Vinnitsa in the Ukraine. Gale performed the transplant on the night of 8 May. Thereafter Piotr lay under intensive care in one of the aseptic units. His skin grew dark, his eyelashes and hair fell out, huge ulcers grew where the beta burns rotted his skin and in the first week of June he went blind.

On 12 June Valeri Perevozchenko died from collapsed lungs and water on the brain. This brought particular sorrow not just to his family but also to Sasha Yuvchenko. He remembered how this former officer in the marines, when they were attempting to lower the control rods, had refused to allow Sasha to look down into the reactor, and in so doing had undoubtedly saved his life. Georgi Popov, the engineer from Karkhov who had come to witness the tests on the turbines, died on the same day from his body's inability to resist infection.

For the next four weeks there were no more deaths, but as Guskova and Baranov knew well, the danger to their patients was not over. Acute radiation sickness follows a well-defined course lasting between two and two-and-a-half months, and until the body has regenerated its own ability to resist disease, a patient can succumb to the slightest infection. Sure enough, on 20 July an operator from Chernobyl died from an ancillary disease; so too did a locksmith from the power station, five days later.

On 30 July it was the turn of Klavdia Luzganova, a sad, middle-aged woman with apparently no relatives or friends, who had worked as a security guard and had never left her post. Her death put an end to her suffering, which had lasted ninety-six days.

Like those who had gone before her, she was buried in the Mitinko cemetery on the outskirts of Moscow. Special concrete slabs were built beneath the coffins of the victims to prevent radionuclides from seeping into the soil from their decomposing bodies.

Luzganova was the last from Chernobyl to die at Hospital No. 6. Together with Shashenok in Pripyat, Lelechenko in Kiev, an unidentified man found dead under a bridge in Pripyat, and Khodemchuk, whose body lay interred in the ruins of the fourth unit, she brought to thirty-one the number of fatalities that could be ascribed directly and unquestionably to the accident at Chernobyl. For the government's medical commission, which had suppressed the news of admissions to other Moscow hospitals and were vague about casualties in Kiev, it became the official statistic. Any higher figure was dismissed as bourgeois falsification and Western propaganda.

To Guskova and Baranov, the disappointment at their failure to save the lives of twenty-seven of their patients was balanced by the recovery of a hundred others. One had a leg amputated, and many still had open sores, but one after another, they were released from the hospital and sent to nursing homes and sanatoriums, either in the Moscow region or in the Baltic states (it was thought important to avoid the strong sunshine in the Crimea). Some went to the sanatoriums of the KGB and were astonished at the luxurious appointments. They were all bald. Arkadi Uskov wore a hat and delayed visiting his mother until his hair had started to grow again. Later in the summer he had to return to Chernobyl to establish his right to receive sick pay. There the trade union gave him compensation for what he had suffered—fifty rubles and an umbrella.

Nina Chugunov, appalled when she had first seen her bald and shriveled husband, now laughed so much at the way his ears stuck out that he felt obliged to wear a beret despite the hot summer sun. He was sent to a sanatorium on the Baltic, but after a couple of days was so bored that he insisted on going back to work. He went first to a non-nuclear power station in Minsk but then, by choice, back to Chernobyl. Four months after he left the hospital, the hair on his head and his body began to grow again.

Most extraordinary—almost miraculous—was the gradual regeneration of bone marrow in both Piotr Palamarchuk and another severely irradiated engineer, Anatoli Tormozin. Each had rejected the donated cells but gradually had recovered the ability to produce blood cells from his own resources. Piotr still had open sores on his legs, back, buttocks, and above all on his shoulders, which had carried the radioactive body of Shashenok. He required further skin grafts and constant medication, but he regained his sight, his healthy skin lost its "nuclear tan" and his eyelashes, fingernails, toenails and hair grew again. Ascribing his recovery to his childhood in a Ukrainian village, this huge, courageous ox of a man pushed back by 180 rems the previously accepted limit of 600 rems for human survival from radiation.

When Razim Davletbayev left the hospital, he went with Inze to the center of Moscow. They thought of all that they had lost in Pripyat—the easy life in a pleasant city set in such beautiful natural surroundings—and knew that it was gone forever. But it seemed insignificant now, when compared to the invisible and intangible good they had encountered in the wake of the disaster—the dedication of the doctors and nurses, the solicitude of their friends, the kindness of complete strangers and the wonder of life itself.

Once again Inze felt compelled to seek some altar at which she could express her gratitude for Razim's recovery, but rather than return to the tomb of Lenin they went to Pushkin Square in the middle of Moscow. There, as they stood in front of the statue to Russia's greatest poet, tears ran down the cheeks of the Tatar engineer. Inze was astonished; she had never before seen Razim cry. He was reading the lines inscribed on the side of the base: "He called on us to be merciful to those in distress."

VIII

1

For Mikhail Gorbachev, the accident at Chernobyl could not have come at a worse time. It was a little over a year since he had been appointed general secretary, and only three months since he had embarked upon the hazardous strategies of *glasnost* and *perestroika* at the Twenty-seventh Congress of the Communist Party. He might have the same title as Leonid Brezhnev or, for that matter, Joseph Stalin, but he had by no means the same power and had gained office by the skin of his teeth. Grigori Romanov, the party boss of Leningrad, had proposed seventy-year-old Victor Grishin, the party boss of Moscow, and it was only at the insistence of the octogenarian foreign minister, Andrei Gromyko, that Gorbachev had been elected.

Once in power, Gorbachev had moved swiftly to prune the dead wood; before the end of the year, Romanov had resigned "for reasons of health," Grishin had been been pensioned off, and Gromyko himself was pushed upstairs as head of state. With them had gone sixteen of the Soviet Union's sixty-four ministers, 20 percent of the party's officials and, in their wake, a legion of dependents in the nomenclatura—a potent source of opposition to Gorbachev's reforms.

It was not these old Brezhnevites, however, who caused Gorbachev concern, but his own ally and second-in-command Yegor Ligachev. A protégé, like Gorbachev, of Yuri Andropov's and brought onto the Central Committee at the same time, he was older than the general secretary and held two posts, which qualified him as a credible rival for power. As the Central Committee secretary responsible for the appointment of party officials since 1983, he could call upon the support of those he had put in power, and as secretary for ideology he was the Grand Inquisitor of communism—both the guardian and the interpreter of the dogma upon which the legitimacy of the government depended.

If the first post had created dependents in the nomenclatura, the second gave Ligachev exceptional influence in a nation still conditioned by the terror of the 1930s, when any ideological deviation led to imprisonment in a gulag or a bullet in the back of the neck. It also suited his puritanical personality. While Gorbachev's style was relaxed, edging toward that of a Western politician, Ligachev retained the severe and formal manner of the party leader. He delivered his speeches in a loud and monotonous tone of voice, but unlike comparable officials from the Brezhnev era, he actually meant what he said. He was just as zealous as Gorbachev and Prime Minister Ryzhkov in the drive for reform, not just of the political habits and economic structures of the country, but of "Soviet Man" himself. Even before Gorbachev came to power, Ligachev had begun the campaign against alcoholism under the slogan "For a sober leadership and a sober population."

While Ligachev may have embodied Communist orthodoxy, he was not alone in being limited by ideological constraints. Every member of the Politburo had been raised under the Soviet system: there was nobody like Slavsky, Alexandrov or Dollezhal who could remember life before the October Revolution. Though those of Gorbachev's generation were only too aware of the problems that beset their country, it remained inconceivable to them that socialism itself could be at fault. If Marx had been God the Father, Lenin was God the Son, the redeemer whose portrait, like an icon, hung on every wall. His word could not be doubted; his church, the party, could not err; and those who had known him, like Efim Slavsky or Armand Hammer, were venerated as the last survivors of the apostolic age.

Equally revered were the heroes of Soviet mythology—men like Kurchatov, the father of the Soviet atom bomb, and his surviving disciples, Alexandrov and Dollezhal. When they had said that Soviet reactors were safe, there had been no reason why Gorbachev or Ligachev, any more than Brukhanov or Dyatlov, should doubt them. Nor had they questioned the blanket of secrecy that had covered the huge state within a state, the Ministry of Medium Machine Building. The rules that had been inherited from an earlier era were the Talmud, if not the Torah; canon law, if not holy writ; and *glasnost*, the policy of openness that they themselves had so recently imposed upon a naturally skeptical people, was meant not to question socialism but to expose such sins against it as sloth, inefficiency, drunkenness and corruption.

For a credible heaven there must be a hell, and the Soviet leaders' conditioning had taught them that evil emanated from the capitalist West. Brought up to believe that their ideological enemies were conspiring to destroy their workers' state, even the younger members of the Politburo were by reflex ready for a fight. Had not the American president, Reagan, described their country as "an evil empire"? Was he not planning a massive escalation of the arms race with his Strategic Defense Initiative? One of Gorbachev's closest allies, Alexander Yakovlev, knew America well. He had been an exchange student at Columbia in the 1960s and had served as ambassador to Canada. Now he was the member of the Politburo who knew most about American foreign policy, and less than a year before had written in the *Literaturnaya Gazeta*, "If the political and military leadership in the U.S. views the balance of power as favorable to such a solution, the Americans do not rule out the direct unleashing of a war against the Soviet Union in their calculations and plans."

There was a gradual but growing realization that even if *glasnost* and *perestroika* succeeded beyond their wildest dreams, even if *Homo sovieticus* sobered up after centuries of tippling and worked as diligently and skilfully as socialist ideology suggested he should, the Soviet Union would be hard-pressed to match the technological pyrotechnics of America's "star wars," but it did not enter the heads of any member of the Central Committee or the Politburo that it was time to throw in the towel and admit defeat. In head and heart, all of them remained loyal to the cause, including the burly giant from Siberia who had replaced the discredited Grishin as party boss of Moscow, Boris Yeltsin. (At the time of the accident at Chernobyl, he was in Hamburg as a fraternal delegate at the congress of the diminutive Communist party of West Germany.)

As a result not just of this conditioning, but of the classified nature of any information about nuclear power, news of the accident at Chernobyl had remained secret. At first the Soviet leaders were informed that it was merely a fire that had been brought under control, and thereafter they were prevented from discovering the gravity of the disaster by the habitual reluctance of lesser officials to be the bearers of bad tidings—or their vain hope that their incompetence could somehow be covered up. It was only when the leaders learned of the furious protests by the Swedes about the cloud of radionuclides that had crossed the Baltic Sea that they appreciated both the gravity of the crisis and its implications for their country's prestige abroad.

On the morning of Monday 28, April Mikhail Gorbachev had called the members of the Politburo to his office for an emergency meeting. Few of them knew what had happened. Geydar Aliev, at one time chairman of the KGB in Azerbaijan and now a deputy prime minister, had heard rumors of the accident but had not been officially informed. Even now the crisis that faced the Politburo was not so much the accident itself as whether or not to admit that it had taken place. The practical and the political were inextricably mixed. The whole population might panic, which would offer an unparalled opportunity in the West for bourgeois falsification and anti-Soviet propaganda. Yet what was the point of a policy of *glasnost* if it fell at this first hurdle?

Different members of the Politburo argued the different points of view. Alexander Yakovlev, head of the propaganda department of the Central Committee, confessed that he did not know what to say. Yegor Ligachev argued forcefully for saying as little as possible. As head of the ideological department of the Central Committee, he had to consider the ideological implications if the full scale of the catastrophe became known. Science was the basis of communism, and communism was the religion of the Soviet state. Like the space program, atomic power was one of their most dramatic achievements. To admit that it had failed—that rather than helping people it had harmed them—would undermine trust in scientific socialism, not just in the Soviet Union but throughout the world.

Geydar Aliev disagreed. "Come off it," he said to Ligachev. "We can't conceal this." The cat was out of the bag. Not just in Sweden but also in Poland and Germany there were reports of unusually high levels of radiation. Soon the whole world would suspect, and it was only by providing truthful, accurate information that they could prevent bourgeois falsification and give *glasnost* some credibility.

Yakovlev came around to Aliev's point of view; it would be best to come clean. But when the vote was taken, Ligachev's views prevailed. A statement was issued throught TASS announcing that the accident had taken place and that a government commission had been appointed. A rider was to be added for Soviet newspapers that nothing was to be reported but this communiqué. The statement was broadcast that night to the Soviet people as the seventh item of news on "Vremya."

Less than an hour later TASS issued a second statement:

> The accident at the Chernobyl atomic power station is the first one in the Soviet Union. Similar accidents happened on several occasions in other countries. In the United States, 2,300 accidents, breakdowns and other faults were registered in 1979 alone, according to the public organization called Critical Mass. The major causes of the dangerous situation are the poor quality of reactors and other types of equipment, unsatisfactory control over the technical conditions of the equipment, non-observance of safety regulations and insufficient training of personnel.

The simple admission that an accident had taken place did nothing to assuage the anger and speculation abroad. An approach from the Kurchatov Institute to the Swedes for advice was leaked to the press and led foreign experts to conclude that there had been either a meltdown at Chernobyl or else an uncontrollable graphite fire. An admission by TASS on 29 April that two people had been killed in the accident was dismissed as improbable; it was reported by UPI on the same date that a source in Kiev "with hospital and rescue team contacts" had said in a telephone interview that "eighty people died immediately and some 2,000 died on the way to hospitals." This figure found its way into the major papers in the U.S. and was reported by Dan Rather on "CBS News." A short time later, UPI reported that Westerners had telephoned their embassies from Kiev to say that up to three thousand people were dead. These rumors gained credence on 30 April when Kenneth Adelman, the director of the U.S. Arms Control and Disarmament Agency, told a Congressional hearing in Washington that two thousand people lived in a village near the reactor, and that the Soviet claims of only two casualties "seemed preposterous . . . in terms of an accident of this magnitude."

A lower estimate of casualties was given by a Dutch radio ham, who had heard from a Russian correspondent that there were "many hundreds of dead and injured and maybe many, many more." In a telephone interview, a Scottish schoolteacher living in Kiev told the London *Times* that local citizens with contacts in the energy ministry had said that up to three hundred people might have died. Plans were made by the British embassy in Moscow to evacuate the seventy British students and tourists in Kiev and the thirty in Minsk. The U.S. embassy advised all of its citizens in Kiev to leave at once. An Austrian firm building a metallurgical plant at Zhlobin, one hundred miles north of Chernobyl, chartered a plane to evacuate the fifty wives and children of their engineers.

The refusal of the Soviet government to release any concrete information made it difficult for responsible Western correspondents in Moscow to refute the rumors that were being relayed back by broadcasts from stations like Radio Liberty in Munich. It was not just professionally frustrating but personally disturbing; the *New York Times* correspondent Serge Schmemann had his wife and three children in Moscow. When it was feared that local produce might be contaminated, the U.S. embassy made its own dosimetric checks on local produce and permitted U.S. journalists and their families to buy supplies brought in from Finland for diplomatic personnel. This not only enraged the Russian government but led to awkward exchanges with Russian friends.

Even among the fraternal Socialist democracies of Eastern Europe there were signs of unease. The Polish government announced restrictions on the sale of milk and warned its people of the absolute necessity of washing all fresh vegetables to remove radioactive particles. Iodine was given to all Polish children under the age of sixteen, and lines formed outside pharmacies in the center of Warsaw.

But there were also more moderate voices in the West. Dr. Thomas Marsham, the director of the northern division of the British Atomic Energy Authority, said he would be surprised if the Soviets did not evacuate people within an area of ten miles, but that he could see good reasons to avoid panic. "You can cause more casualties if you panic. If you were to suddenly say everyone should get out of Kiev, you'd probably have two hundred people killed on the roads. But there's a difference," he added, "between causing panic and telling people what's going on."

· · ·

Although they were not privy to the proceedings of the Politburo on 28 April, it was quite clear to Western journalists and diplomats what was going on in government circles. "To the world outside," wrote Schmemann in the *New York Times,*

> almost as striking as the nuclear accident that sent radioactive debris over hundreds of miles was the Soviet effort to restrict information about it.
>
> It was a reflexive retreat into secrecy that once again seemed to show the Kremlin loath to concede the smallest failing before its people or a hostile world. . . . The reaction spoke of a compulsion to maintain face. . . . Diplomats agreed that the nuclear accident was a major embarrassment for Mikhail S. Gorbachev and the Kremlin, at a time when the Soviet Union is talking of a "new style," of more openness, of less centralization, of more individual responsibility, of more candor about failings.
>
> It undermined the image cultivated by the Kremlin as the globally recognized champion of a nuclear-free world, opposed by a militaristic Washington. It cast a blot on Soviet technology, in an area in which officials have always insisted that the Soviet Union has no equal. It stood to put Moscow on the defensive on environmental protection, which has been given prominence under Mr. Gorbachev.
>
> Yet what seemed paramount in the reluctance to release information was the Soviet conviction that a hostile world was ready to use any ammunition available to attack the Soviet state.

2

Deciding that the best defense was to counterattack, on the very day, 2 May, that this analysis by Schmemann appeared in the *New York Times,* Academician Alexandrov called a meeting of senior scientists in his office in the Kurchatov Institute. Among them was Yuri Sivintsev, a dapper, fast-talking expert on nuclear safety who had been brought to the institute by Kurchatov himself in 1948. He knew about the accident and had reason to believe that it was serious, having heard that the existing patients had been evacuated from the nearby Hospital No. 6 to make way for the casualties.

Sivintsev found Velikhov with Alexandrov, and assumed at first that he had been summoned to advise on the accident itself; instead, he was asked by Alexandrov about accidents in nuclear power stations abroad. He told him about the fire at Windscale in Britain and the accident at Three Mile Island in the United States. He was told nothing about what had happened at Chernobyl, but assumed the accident there was on a similar scale.

The accidents at Western reactors were now repeatedly cited in the Soviet press, while the Western reaction to Chernobyl was presented as a deliberate attempt to escalate the Cold War. Wrote Yuri Zhukhov, a commentator, in *Pravda:*

> The United States leaps at any excuse to heat up further an already tense situation, sow distrust and discord among the peoples, and poison the political climate. And the purpose of it all is to divert attention from the criminal aggressive actions of the United States, like the recent bombing of Libya and the undeclared wars against Afghanistan, Angola and Nicaragua, to justify the intensification of the arms race, the continuation of nuclear tests, and the refusal to accept Soviet peace initiatives.

Zhukhov charged that the escalation of these provocative actions had reached a new peak as "the U.S. state apparatus and the news media which do its bidding put out some fabrications about the consequences of the accident at the Chernobyl AES."

> It was only senior figures in Washington and in the capitals of some other NATO states who immediately latched on to the news of the Chernobyl accident in order to exploit it for their own hostile political ends. The fueling of hysteria and panic began. Cock-and-bull stories were concocted about "thousands of dead" and about the possibility of the population of Western Europe and, in all likelihood, the United States being affected by radioactivity.

Prompted by Washington's special propaganda services, the West European gutter press started concocting fabrications, each one more awful than the one before. Expert panic-mongers began the forced evacuation of Western students, specialists and tourists from the USSR, even if they were in Siberia.

The same line was taken in Soviet broadcasts on radio and television, and in other Soviet journals, such as *Izvestia* and *Literaturnaya Gazeta.* In Hamburg Boris Yeltsin had been giving a reasonably accurate account of the accident, denouncing reports of widespread damage to crops and the contamination of water and milk throughout the Ukraine:

This whole propaganda campaign can only arouse indignation. This campaign follows political motives. And you can note that this campaign is carried on in countries whose governments stubbornly refuse to cooperate in the program of abolishing nuclear arms and means of mass destruction.

At the very moment when the medical commission was ordering checks for radioactive contamination of all passengers traveling on trains passing through the Ukraine, TASS was denouncing the "humiliating examination" for radioactive contamination of a troupe of Ukrainian actors at a border crossing from East into West Germany:

Unleashing a campaign of this kind was promoted by the FRG official authorities' stand, who wittingly gave distorted information on the consequences of the accident at the Chernobyl AES. The extremely right-wing forces in the FRG, gambling on this misfortune, were trying to make use of it to extirpate the shoots of détente appearing in Europe.

The counterattack failed. Despite the gruesome exaggerations in the Western press, it was impossible to sustain the outrage and at the same time retain the old policy of secrecy, which the Politburo had adopted at Ligachev's insistence. It was coming under strain both at home and abroad. The governments of the Scandinavian countries made repeated demands for further information, and the Politburo's own medical

commission asked that the facts about radiation be broadcast on television. There was evidence of growing panic. After the discovery that
several trains had been contaminated, dosimetric controls had been
ordered by the government's medical commission on all passengers
arriving from Kiev, Gomel and Minsk. This had caused such alarm that
the order had been rescinded.

There was further evidence of panic in the lines formed at pharmacies
for iodine prophylactics. Initially the commission had accepted Professor
Vorobyov's assurance on 2 May that it was too late for stable iodine to
provide protection, and offers of supplies from abroad were turned
down. However, Ilyn in Kiev sent a coded telex to Mikhail Gorbachev
warning him to expect serious consequences in the middle future from
the radioactive iodine that had accumulated in the thyroids of children,
and when, on 6 May, it was reported that "the level of radiation did
require iodine prophylaxis," the commission changed its mind. It was
estimated that 218 tons would be required in the Russian federation
alone, and a request was made to the Ministry of Chemical Production
to convert a factory for its production. The commission vacillated about
who should take stable iodine. The Soviet ambassador in Romania was
told firmly that no Soviet diplomats were to accept the iodine being
offered in Bucharest, but while one member of the commission suggested that tests on patients in Hospital No. 7 had shown that only the
citizens of Pripyat had received significant doses, another reminded his
colleagues that whether or not the iodine prophylactic was needed,
doctors must respond to the appeals of any citizen for whatever reason,
in particular for medical assistance. People's psyches were disturbed and
doctors had no right to refuse them treatment, even if for all practical
purposes it was of no use.

It was the same when it came to information. The first instinct of the
commission was to keep all data secret, and a request was sent to a
Comrade Zorin to make sure that no information about the medical
aspects of the crisis be published in the press without its consent.
Information about the condition of patients in Hospital No. 6 could be
released because Gale and his friends would have it anyway, but none
was to be released from Hospitals No. 7 and 12.

On the other hand, with the evidence of growing panic, the medical
commission made a request to Ryzhkov's Politburo commission for
more information to be broadcast on television to reassure the popula-

tion. With this advice, and a growing clamor for more information from abroad, the uncompromising line favored by Ligachev was modified by two concessions. First, the director general of the International Atomic Energy Agency in Vienna, Hans Blix, was invited to inspect the ruined reactor at Chernobyl together with his head of atomic safety, an American named Morris Rosen; second, a televised press conference was called for 6 May at the Ministry of Foreign Affairs.

That morning stories appeared in *Pravda* and were put out over the wires of the Soviet news agencies describing, with some accuracy, the accident and the measures being taken to monitor radioactive contamination. The same relative candor was evident at the press conference in the briefing hall of the Ministry of Foreign Affairs. It was opened by the deputy foreign minister, Anatoly Kovalev, and then addressed by the chairman of the first government commission, Boris Scherbina himself.

It was a relative novelty for Soviet officials to hold Western-style press conferences of this kind, and its proceedings were strictly controlled. The room was packed not just with reporters from many nations, but also with foreign diplomats, including several ambassadors, and the spectacle was filmed by more than a dozen television cameras. However, questions were only taken from Soviet reporters or foreign correspondents from Communist papers, and they were paraphrased to suit the prepared replies.

Scherbina gave an account of the accident that, though it passed over the graver aspects of the crisis and gave a misleading impression of the competence of his commission, at least contained no outright lies. His estimate of casualties was based on the initially optimistic figures given by Baranov and Guskova. He suggested that the accident was being exploited by Cold Warriors in the West to frustrate the peace initiatives of Mikhail Gorbachev.

When asked by foreign journalists whether there was any danger to the population of other countries from the fallout from Chernobyl, Y. S. Sedunov, Yuri Israel's deputy on the State Committee of Hydrometeorology, said, "In our opinion, there was no direct threat to the population either in our areas that are far away from the site of emission, or of foreign countries." He admitted enhanced levels of radiation in neighboring countries, but felt that "this emission is shortlived, insignificant, not high."

To a follow-up question about levels of radioactivity, Sedunov painted a reassuring picture by carefully selecting the statistics he chose to reveal. In Moscow there was no change to the background level. In Chernobyl the level had risen to fifteen milliroentgens an hour but had already dropped threefold. In Kiev the level at the time of the explosion did not exceed background levels, and three days ago was 0.2 milliroentgens an hour. In Minsk there was no substantial increase over the natural background level. In Gomel the situation was approximately the same as in Kiev.

There were further questions about the safety of Soviet nuclear power stations, which Petrosyants defended with vigor. Forty-one units had been working safely for the past thirty years. The technology was young, complex, and in a number of ways problematic, but other countries had faced similar problems. Look at Windscale, Idaho and Three Mile Island!

After only an hour, with the hands of many Western reporters still raised to ask questions, the conference was brought abruptly to an end. However, excerpts from the news conference were shown on television, both on "Vremya" and in a special bulletin that followed. All references to the possible dangers from radiation had been removed.

3

In the days that followed, the different approaches of Ligachev, the Central Committee secretary for ideology, and Yakovlev, the secretary for propaganda, were apparent in the tone taken by journalists loyal to different masters. On 7 May *Sovetskaya Rossiya* mentioned merely a "lack of coordination" in the evacuation of the population of the contaminated area; on 12 May *Pravda* named officials guilty of grave dereliction of duty. Sometimes contradictory stances even appeared in the same paper. In a front-page article on 18 May, *Pravda* criticized those who refused to take the Soviet people into their confidence; on 21 May this was contradicted by a second article saying that criticism in the West of the dissemination of information was totally unjustified, since collecting accurate data inevitably takes time.

This battle was not so much about the aftermath of the accident as over what meaning was to be given to *glasnost*. Gorbachev had launched the policy, but seemed uncertain about how far it should be allowed to go. In theory he saw the need for it, but as a Soviet patriot he was loath to wash his country's dirty linen in public, and as he revealed to Armand Hammer and Robert Gale when he received them in the Kremlin, he was genuinely angry about the way the Western media had gloated over this misfortune.

Throughout the party apparatus, and particularly in the media, the wily watched and waited to see which way things would go. Both Ligachev and Yakovlev had their supporters, and cautious editors hedged their bets by giving space to each. In *Pravda,* commentators like Yuri Zhukhov hammered away at the wickedness of the West, but Yakovlev had both a friend and an ally in the science editor, Vladimir Gubarev. A balding, handsome man, he had the kind of tough, chiseled face given to "positive" Soviet heroes in paintings of the Stalinist era. With two degrees, one in physics from Moscow University, the other from the Institute of Construction and Development, Gubarev was well qualified as a scientist. But he was also the director of television films and the author of half a dozen plays and two dozen books, several of which had been published abroad. In the company of journalists he said he was an artist; in the company of artists he said he was a journalist. In both roles, he was a Communist, and in over thirty years at *Pravda* had grown adept at following the different twists and turns of the party line.

On 26 April Gubarev had been tipped off about the accident by a physicist from the Kurchatov Institute. He had immediately telephoned the *Pravda* correspondent in Kiev, Mikhail Odinets, and told him to investigate. Odinets rang back later that day and said that there had been roadblocks on the way to Chernobyl; the militia had let him through, but later the KGB had sent him back. Three days later, on 29 April, after the TASS communiqué about the accident, Gubarev went to see his editor with his suspicion that the accident must have been much more serious than they had been led to believe. When the editor telephoned the Central Committee, he was told to drop the matter and not to interfere. This confirmed Gubarev's suspicions. He himself telephoned Yakovlev, but his friend, bound by the decision of the Politburo, fobbed him off. "I'm told it's nothing serious. Don't get involved."

On 2 May, after Ryzhkov and Ligachev had returned from Cherno-byl, six trusted journalists were authorized by the Politburo to visit the scene. With the local correspondent, Odinets, Gubarev boarded a hy-drofoil that was to take samples of water where the Pripyat River entered the Kiev reservoir.

In a leisurely, almost literary dispatch that was later published in *Pravda,* Gubarev and Odinets described their journey. The article was filled with facts: the Kiev reservoir was a man-made sea occupying an area in excess of 920 square kilometers and containing 3.7 cubic kilometers of water. The Polesskoe and Ivankov regions had sown over 650,000 hectares of crops that spring, which was 100,000 more than the year before, including 150,000 of corn, which was 40,000 hectares more than the year before. Chernobyl was the repair base of the Dnieper Steamship Company as well as a center for pig-iron production, cheese and mixed-feed factories, with an industrial combine, a consumer-ser-vices combine, medical and agricultural vocational and technical col-leges, four general education schools, a musical school, a hospital, a polyclinic, a culture center, a cinema and a library.

There were uplifting interviews. "We people belong to one family," said an Estonian engineer. "If the Ukrainian city of Chernobyl has suffered a misfortune, we are ready to come to its assistance." "Soviet people have the same joys and the same concerns," said another, from Kirghizia. "We have a sacred principle," said a third, from Belorussia. "The sacred law of brotherhood."

At the headquarters of the commission in Chernobyl, Gubarev and Odinets interviewed Velikhov, who described the efficiency with which they were dealing with the crisis. "Previously it took months to reach agreement, but now a night is enough to decide virtually any problem. There is not a single person who refuses to work. Everyone is acting selflessly." He conceded that the problem they faced was unusual and "requires solutions that neither scientists nor specialists have ever had to deal with before," but the main lesson to be learned was "how cata-strophic nuclear war is." The accident was trifling by comparison. "Over there in the West, particularly in Europe, people are shouting and making a din about Chernobyl, but they themselves are keeping quiet about or are trying to belittle the danger that the Pershings' nuclear charge contains. So it is worthwhile for Western propagandists to con-sider: should they gloat over an accident that has occurred or would it be better to prevent a worldwide catastrophe?"

Despite its prosaic presentation and ideological dressing, Gubarev's article gave several indications to the careful reader of the gravity of the accident at Chernobyl. He did not interview Legasov, who was loyal to Ligachev, head of the department in which his father had served in the secretariat of the Central Committee. Velikhov, however, was willing to give substance to *glasnost*, and for Yakovlev, Gubarev's visit presented an opportunity to present information to Gorbachev that might not be reaching him through official channels. When Gubarev returned to Moscow, he was summoned to the Kremlin by Yakovlev, who then took him to see Gorbachev in his study. There Gubarev told the party leader of the chaos and incompetence he had witnessed, and warned him that the aftermath of the accident would be with them for many years to come. The general secretary seemed taken aback at the pessimism of his account.

4

The information about the accident, cautiously and selectively imparted through the Soviet media, did little to reassure the citizens of Kiev. Clips of film showing cows grazing in the shadow of the reactor, which had been broadcast on television in the days that followed the accident, were now revealed as lies to a public already prone to distrust any form of official propaganda. They were more willing to believe the news that filtered in from Radio Liberty, the BBC or simply telephone calls from friends abroad, that Western governments were repatriating their citizens from all parts of the Soviet Union.

Controlled by the system of internal passports and unable to leave their jobs without good reason, workers in offices and factories in Kiev now put in for their vacations, paid or unpaid, and when these were refused even proffered their resignations. On the night of 5 May, people slept at the railway station to keep their place in the line for tickets. By the next morning the ticket offices came under siege, and soon the only seats available were on planes and trains departing five or six days later. Pensioners and invalids from the Great Patriotic War, who had precedence, were brutally elbowed aside as parents scrambled to get tickets for their children. A brisk black market sprang up, and tickets were sold for a premium of fifty or one hundred rubles.

On 6 May, ten days after the accident, at the same time as the first descriptions of it were published in the press, local radio and television stations broadcast the first advice to the inhabitants of Kiev by the Ukrainian health minister, Anatoli Romanenko. This was to wash their vegetables, close their windows and remain indoors. Many preferred to try to leave. Rumors began to circulate about how the party leaders had sent their children and grandchildren to camps and sanatoriums in the Crimea as soon as the accident had occurred. Crowds formed at the banks, and some banks were forced to close only an hour or two after they had opened. Others limited withdrawals to one hundred rubles, but even that depleted their deposits, and by the afternoon of 6 May the banks in Kiev closed because they had run out of money. Those people with cars now tried to flee, and traffic jammed the southern routes out of the city.

Those who remained, hearing that iodine was a precaution against radiation, flocked to pharmacies, but found that the stocks had sold out. Some got hold of iodine intended for external application, drank it down and scalded their throats. The other reputed antidote was vodka, and the usual lines outside the liquor stores quadrupled in size. The short opening hours decreed on 1 May as part of the drive against alcoholism were abandoned. Believing themselves protected by their intoxication, drunks ventured out in the street. A group tried to pick up a disheveled young woman in the Khreshchatyk boulevard; it was Lubov Kovalevskaya, the journalist from Pripyat, who had just returned from the airport where she had sent her daughter off to stay with her sister in Sverdlovsk.

Since she had to pay a black-market price for the ticket for her daughter, Lubov had returned to Kiev with nowhere to stay and only twenty kopeks in her purse. She was in a state of shock, crazed, unkempt and dirty. She tried to recite some lines of her own poetry, but none came to mind. She could not even tell left from right. She was standing in line for a taxi without any clear idea of where she should go when a man asked if she was from Chernobyl. When she said she was, he took her by the arm, gave her supper at his office, then arranged for a room at the Hotel Moscow. She remained there for five days at his expense; he asked for nothing in return.

· · ·

The anxiety and exasperation of the inhabitants of Kiev were shared by their leaders. Vladimir Shcherbitsky, Brezhnev's old crony, who was still general secretary of the Ukrainian Communist party, had done everything that had been expected of him. Minor officials might have sent their families out of the city, but he had stood on the rostrum on May Day with his grandson at his side and had appeared at the international bicycle race that had passed through the city on 9 May. But since the Chernobyl station was an All-Union enterprise, and since anything to do with it was kept secret from the provinces' leaders, he knew little more of the true situation than the man in the street.

His minister of health, Romanenko, was equally ill informed. Returning from a conference in Atlanta, Georgia, on 2 May, he had gone to Chernobyl on the 3rd and gained some idea of the seriousness of the accident, but all authority had been vested in the government's medical commission in Moscow, headed by the Soviet minister, Shepin. Romanenko's task was to carry out its instructions as best he could, and this he did by mobilizing the entire health service of the Ukraine. He formed a thousand additional medical teams to treat the evacuees from Pripyat and the thirty-kilometer zone, and to examine the inhabitants of villages on the fringe. This put an enormous strain on their resources since there were not enough doctors or laboratory technicians to make blood tests on so many thousands of people. Eventually the army medical corps was mobilized to assist them, but the criteria upon which their care was based was stipulated by the Institute of Biophysics under Academician Ilyn and the State Committee of Hydrometeorology under Professor Yuri Israel.

Knowing as little about radiation as anyone else, the Ukraine's political leaders became infected by the growing panic in Kiev, and on 7 May they summoned both Ilyn and Israel to a meeting of the Politburo of the Central Committee of the Communist Party of the Ukraine. There Shcherbitsky insisted that they draw up in writing their recommendations for the protection of his people from the fallout. It was an unusual request, suggesting an ominous mistrust of the central organs of power. However, Ilyn and Israel agreed, and retired to prepare the paper.

The Ukrainian Politburo reconvened at 11:30 P.M. The party leaders, including Shcherbitsky and the redoubtable Ukrainian minister Valentina Shevchenko, sat facing Ilyn and Israel across the wide conference table. They presented their report as follows:

1. The radiation situation in the city of Kiev and the oblast at present does not present a danger to the health of the population, including children, and is within the limits of norms recommended by national and international atomic-energy agencies in case of accidents at a nuclear power station.
2. The level of radioactive substances contained in food products at present presents no danger to the populations.

Ilyn's and Israel's recommendations were:

1. An analysis of the radioactive situation in the city of Kiev demonstrates the absence, at present, of indications of the need for the population, in particular children, to be evacuated to other regions.
2. The population should be informed through the mass media, television and radio about the situation and the measures being taken.

This was signed by both Israel and Ilyn and presented to Shcherbitsky, but before the meeting broke up Valentina Shevchenko suddenly leaned forward and asked Israel, "Yuri Antoniyevich, what would *you* do if your own grandchildren were in Kiev?"

She was later to claim that Israel made no answer, while Ilyn insisted that Israel had reassured her. Whatever the truth, a resolution was passed the next day by the Ukrainian Council of Ministers, with Israel's concurrence, bringing the school year to an end on 15 May and evacuating children up to the seventh grade from Kiev to areas unaffected by the fallout from Chernobyl.

On the evening of 8 May, Romanenko spoke on Ukrainian television. He reassured the viewers that since his earlier broadcast the radiation situation had improved. The level of background radiation was gradually falling and was well within the norms recommended by national and international bodies and did not represent a danger to the health of the population, including children. Nevertheless he drew attention to the danger from radioactive dust. The intensive washing of the streets and hosing down of apartment blocks would continue, and people were advised to take a shower and wash their hair every day. Children could play outdoors, but only for a limited period of time. "They want to be out in the open air, so let them play, but not as is usual in good weather from morning until night, but just for the odd hour, and they should not kick balls around in dusty areas."

Romanenko then announced that "to improve the health of the children of Kiev city and oblast" the school term would end on 15 May. The children would be sent to Pioneer and holiday camps in the southern oblasts "in a well-organized manner, and for this purpose an adequate number of trains and vehicles are being set aside." In conclusion, he stressed, "All matters connected with the influence of the environment on the health of the population are constantly being monitored by the Ukrainian Republican Ministry of Health."

5

These measures by the Ukrainians were not approved by the government's medical commission. The dose established by Ilyn's Institute of Biophysics as virtually harmless was twenty-five rems. This was the special exposure limit set for members of the armed forces dealing with the aftermath of the accident at Chernobyl. On the very day of Romanenko's broadcast, the commission proposed to set a limit five times lower, at five rems, which would mean that at the current level of contamination, the inhabitants of Kiev were safe for two and a half years. Professor Vorobyov confirmed that this left a vast margin of safety. On 5 May he had been anxious when it was reported that the medical personnel in Hospital No. 6 had been receiving over a short period doses of more than ten to fifteen rems, and had ordered them to work in shifts. But on 8 May, in considering emergency-dose levels leading to intervention, he asked the commission to look at the facts, which were that even one hundred rems did not lead to leukemia. The chromosomes repaired themselves. A favorable prognosis could be established for one generation. As the deputy minister of health, Shepin, reported to the medical commission, the decision of the Ukrainian leadership to evacuate children from Kiev was "an emotional reaction without any objective justification." It was also the first intimation that the scientists were no longer in control.

As reports came in of contamination far beyond the thirty-kilometer zone, including vast tracts northeast of Gomel in Belorussia, the medical commission gave orders for the evacuation of the inhabitants of some affected areas and the provision of uncontaminated food and milk in others. There was considerable confusion in the field about the dose levels that should trigger this kind of government intervention. The commission was told that medical staff in the field were applying the ten-rem limit, drawn up to limit their own short-term exposure, as the intervention level for the estimated lifetime exposure of the general population.

After repeated requests to the commission to arrive at definitive norms, a working party of experts was set up and on 12 May it recommended that temporary norms of ten rems of external radiation to the whole body and of thirty rads to the thyroid be set for children and pregnant women. For the rest of the adult population, intervention was to take place when the estimated lifetime dose was fifty rems. This was approved by the commission for a month, after which it was was reduced to thirty-five rems.

Although areas to the northeast of Gomel were evacuated on 2 May, and measures taken to provide those who remained with "clean" imported milk and food, the contamination of about fifteen hundred square kilometers of land so far from Chernobyl was not mentioned on television or in the press. The public was left with the impression that pollution was restricted to the thirty-kilometer zone. There was a fear of widespread panic if it were to become known that land three hundred kilometers north of Chernobyl had been rendered uninhabitable by the fallout, or that significant deposits of radioactive cesium had been found five hundred kilometers to the west, only two hundred kilometers from Moscow.

There was also the preoccupation with Soviet prestige. It was thought that reducing the scale of the disaster would limit the damage done to the country's reputation abroad. For the same reason it also became a matter of urgency for the government to come up with an ideologically acceptable explanation of how the catastrophe had occurred.

6

On hearing of the accident at Chernobyl on 26 April, the public prosecutor in Kiev had opened an investigation into a possible offense under Article 220, Section 2, of the criminal code, which covered breaches of regulations at "explosive-prone enterprises or in explosive-prone workshops." The file was tagged as Criminal Case No. 19-73. The following day, 27 April, a team was formed to conduct an investigation, which included not just lawyers from the prosecutor's office, but investigators from the Ministry of the Interior and the KGB. At the same time, a parallel investigation was started in Moscow by the deputy general prosecutor of the USSR.

The investigators, however, worked under two constraints: the first was the subordination of all who worked in the Soviet judicial system to the dictates of the party; the second was the abstruse nature of the evidence, which only a few physicists could understand.

Immediately after the accident, an investigation into its causes was set up by the different ministries and institutions involved in atomic energy—the Ministry of Medium Machine Building; the Ministry of Energy and Electrification; Alexandrov's Kurchatov Institute; Dollezhal's design institute, NIKYET; Abagyan's institute, VNIIAES; the State Committee for the Use of Atomic Energy; Hydroprojekt; and by the chief engineer of Chernobyl, Nikolai Fomin. Reporting to the commission on 5 May, this committee of inquiry pointed out that the program drawn up by Dyatlov for tests of the turbines was flawed; that in implementing it the operators had broken certain safety regulations and the accepted technical procedures; and that the reactor had gone out of control because of these mistakes. However, it recommended that the two ministries running RBMK reactors should reconsider whether in fact they were complying with certain laws on industrial safety, and if it transpired that they were not, should take strict measures to comply with them in all operating RBMK power stations and those under construction.

At the same time, on 26 April, a second investigation was started by Boris Scherbina, head of the government commission. This was led by Armen Abagyan, chief of VNIIAES, the All-Union Research Institute for Nuclear Power Plant Operation, and included representatives from the same ministries and institutes, among them Alexander Kalugin from the Kurchatov Institute, as well as senior figures from the prosecutor's office, the Interior Ministry and the KGB. The legality of this investigation's proceedings was supervised by the deputy general prosecutor of the USSR, O. V. Soroka, and its final report was signed by the general prosecutor of the USSR, A. Reykunkov.

This committee also blamed the Chernobyl personnel for the breach of safety regulations, including not just Dyatlov as the deputy chief engineer responsible for the tests, but also Brukhanov and Fomin, who "made grave errors in the running of the nuclear power station and did not ensure its safety." These errors were regarded "above all else" as the real cause of the accident at reactor No. 4.

However, some additional blame was attached to the Ministry of Energy and Electrification for allowing such tests to be carried out at night and for tolerating "physical and technical shortcomings in the RBMK reactor" without insisting that the chief designer and scientific leader take the necessary measures to improve the safety of the reactor. There was also criticism of the Ministry of Medium Machine Building, which had permitted the same shortcomings in its RBMK reactors, and, in brackets, as if in passing, its leader, Efim Slavsky, as well as Alexandrov, the scientific leader; Nikolai Dollezhal, the chief designer; and Ivan Yemelyanov, his deputy at NIKYET. All of these were rebuked for failing to implement timely measures to improve the safety of the RBMK reactors in accordance with the requirements "stated in general safety regulations" for nuclear reactors at any stage in their design, development or exploitation. "In the design of the reactor," the report concluded, "there were insufficient technical measures to ensure its safety."

The ripples of responsibility for the accident spread further still. Criticism was directed at the State Committee for Safety in the Atomic Power Industry, which "did not ensure adequate supervision of the implementation of norms and regulations concerning nuclear safety, and did not make full use of its statutory rights." Its leaders, Kulov and Sidorenko, had been indecisive and had not stopped "the breach of norms and safety regulations by the various ministries, enterprises, power stations and factories that supplied equipment and devices."

Though mistakes made by the operators initiated the explosion, the report went on, the explosion itself stemmed from flaws in the reactor's design that permitted a positive void coefficient in the core and the possibility of a sudden surge of power. There were no systems incorporated in the design to deal with such eventualities, and there was a fatal flaw in the specifications for the control rods, which had led to positive reactivity in the first moments of their descent into the active zone. The design of the reactor did not include a device that would show the operating reactivity or threshold of danger.

Although it ascribed some culpability not just to the designers but also to those who knew of the potential hazard, this report to the commission did not make clear that there was no mention of any danger in the operating manual of the RBMK reactors. An experienced operator should have realized that it was dangerous to remove so many control rods from the core of the reactor, but not that pressing the emergency button to return them would start the process that had led to the explosion. The report made much of the breach of safety regulations, but it did not say that operators were permitted to disregard them on the authority of a deputy chief engineer—in other words, that these regulations were more like guidelines than laws; nor did it point out that most of the breaches, such as disconnecting the emergency core-cooling system, had no bearing whatsoever on what had gone wrong.

Perhaps all this was only to be expected from a team of investigators who were in essence investigating their own failings. Since there were representatives from the Kurchatov Institute and Dollezhal's design bureau on the investigating team, but none from the personnel on the spot, it was only to be expected that the emphasis should be put on the shortcomings of the operators, and the word went out almost at once that their reckless mistakes had caused the disaster. If the design of the reactor itself was blamed, the pressure of public opinion both at home and abroad might force the closing of all the other RBMK reactors, with catastrophic consequences for the Soviet economy.

Certainly the fallibility of the operators was the easiest face-saving explanation to feed to the bourgeois press in the West. As early as 30 April, Boris Yeltsin in Hamburg told a West German television network that the accident had been caused "by human error," and on 4 May he told the Associated Press that "one thing, certainly, is hard to believe, and that is that it had anything to do with the quality of the equipment."

At his press conference on 6 May, Boris Scherbina announced that a full investigation of the causes of the accident was under way:

> ... but taking into account that the design and structural solutions correspond fully to the norms of both our country and generally accepted international practice, and that the quality of the manufacture, the installation and acceptance of equipment was properly checked, the cause of the accident could be the consequence of the coincidence of several exceptionally unlikely and therefore unforeseen failures. The activity of the staff on duty is also being analyzed carefully.

When he addressed the nation on television on 15 May, Mikhail Gorbachev refrained from saying what had gone wrong. "It is as yet too early to pass final judgment on the causes of the accident. All aspects of the problem . . . are under the close scrutiny of the government commission."

Equally reticent were the two officials from the International Atomic Energy Agency in Vienna: its Swedish director general, Hans Blix, and the head of its nuclear safety department, the American Morris Rosen. Both had flown over the reactor in a helicopter on 7 May, and on 9 May held a press conference at the Ministry of Foreign Affairs in Moscow. They confirmed that "a full and authoritative description of the accident, the causes of the accident, and its consequences, can only be given by the Soviet authorities after the necessary analysis." In their own summary description, they reassured the journalists that "as a prophylactic measure, people inside and outside the thirty-kilometer zone were given potassium iodide tablets." Rosen assured them that "the accident area will be inhabitable again," and in a reply to a suggestion from the Novosti Press Agency that the Western media were trying to frighten the public, Blix asked the reporters present "not to exaggerate, not to disseminate rumors of different kinds or any kind of information that will only worry people."

"Mr. Director General," asked a correspondent from *Pravda*, "are you satisfied with the volume and the nature of the information that you have received during your visit to the USSR?"

"To this question I reply unquestionably yes," replied Blix. "We had very frank and open discussions with ministers and experts, and in many cases the experts and officials are people whom we have known for a long time."

Before returning to Vienna, Blix and Rosen were promised by the Soviet authorities that a full explanation of what had happened would be presented to the IAEA in due course. Back in Vienna, they were subjected to a more rigorous interrogation by two correspondents of the German magazine *Der Spiegel*. "The Russians are confident," said Blix, "that they will be capable of decontaminating the area. It will be agriculturally usable again."

"How soon?"

"We did not talk about when they will begin or how long it will take."

"How strong was the radiation intensity—four hundred or even one thousand rems?"

"We did not ask," said Rosen.

"Why not?"

"We did not go there to ascertain what dose of radiation has hit the populace."

"It is hardly comprehensible that you did not raise this question, since it is highly significant for the consequences in all neighboring countries."

Later, the two interrogators from *Der Spiegel* asked Rosen if he had raised the question of the RBMK's unsatisfactory safety standards before the accident.

"We do not have any proof that they are faulty," Rosen replied.

"Can you tell whether the Soviet reactors are safer, or less safe, than the reactors in the West?"

"They are a different type."

"A month prior to the accident, Lubov Kovalevskaya, obviously an insider, published in the journal *Literaturnaya Ukraina* a report on the Chernobyl power plant. It constitutes a horror picture of mishaps that occurred during the construction of the reactor—slovenly work, decrepit material, dangerous haste. Did you read the report prior to the accident?"

"No," replied Blix. "I did not read it."

"I know some of the commentaries made in that respect," said Rosen. "There are similar reports on slovenliness in the construction of U.S. reactors."

"This report did not make you nervous?"

"Nervous? Why?"

"Because of the dangers to mankind and the environment . . . "

"I am interested in commentaries on all reactors," said Rosen. "I file them away for potential future reference."

Der Spiegel suggested that perhaps there should be an organization like Amnesty International to monitor the dangers inherent in nuclear energy. "If we take a look at what you publish," Blix was told, "we gain the impression that your organization is nothing more than a public relations agency for the application of nuclear energy."

"No, quite the contrary," said Blix. "You are quite wrong there."

7

The world now awaited the full disclosure of the causes of the accident, which Gorbachev himself had promised would be made at a conference in August of the International Atomic Energy Agency in Vienna. The task of preparing the report was entrusted to a group of twenty-three experts—half of whom were from the Kurchatov Institute, including Legasov, Ryzantzev, Kalugin and Sivintsev; the other members included Ilyn, Israel, Abagyan, Guskova and Pikalov. Papers were also submitted by the very institutions that had an interest in concealing the truth: the Kurchatov; NIKYET, Dollezhal's design bureau; Zukh-Hydroprojekt, which had built the reactor; Abagyan's research institute, VNIIAES; and, protecting them all in its huge embrace, Slavsky's Ministry of Medium Machine Building.

It was not merely a question of saving the reputations of a few physicists; they could have been thrown to the wolves. There was the far greater danger that if it was admitted that RBMK reactors were faulty by design, responsibility for the accident would reach up to the minister, Slavsky, to the chief designer, Dollezhal, and to the scientific leader and president of the Academy of Sciences, Anatoli Alexandrov himself. This was unthinkable. To the generation of political leaders now in the Politburo, these three octogenarians were like living statues in the pantheon of Soviet heroes. To pull them down, to show the world that they were careless, incompetent or simply mistaken, would fatally discredit the whole Soviet system. It would demoralize the scientific community; in their own institutes, they were revered, and the accident at Chernobyl was seen as a tragedy not merely for the harm it had inflicted on its victims but for the blight it had cast on their life's work.

No one felt more protective of Alexandrov than his first deputy director, Valeri Legasov; yet no one knew better the full extent of the catastrophe. Before the accident he had been nominally responsible, along with Alexandrov, for all atomic enterprises in the Soviet Union. Together with Alexandrov, he had insisted that Soviet reactors were safe.

On 26 April snatched from a party cell meeting at the Ministry of Medium Machine Building, as a member of the government Legasov had set off for Chernobyl in his smart suit and leather coat, imagining a mishap of the kind the Americans had suffered at Three Mile Island. Ten days later, on 5 May, he had returned briefly to Moscow in the uniform of a soldier. His suit and leather coat were both highly radioactive. "If you can decontaminate these," he said to his friend Protzenko, "you'll win the Nobel prize." He was physically exhausted, but also morally shaken by what he had witnessed in the past two weeks. "Such unreadiness," he told another friend, "such disorder. A worst possible version of 1941 with the same courage, the same desperation and the same unpreparedness . . . " With tears in his eyes, he told his friends how everyone had been taken by surprise. There had been no stable iodine and none could be obtained; nor were there any respirators, medicines, reserves of fresh water or uncontaminated food. He feared the accident's long-term effects.

Legasov returned to Chernobyl to lead, with Velikhov, the struggle to contain the consequences of the disaster. The prodigious power of the Soviet state, marshaled by Ryzhkov—tens of thousands of troops, fleets of helicopters, battalions of miners, metro construction workers and engineers—looked for guidance to the handful of physicists to tell them what action was necessary. Some, like Kalugin and Fedulenko, might know more about the RBMK reactor than more eminent academicians, like Velikhov and Legasov, but because they were not leaders and did not have the stature to share in the deliberations of the commission, they were ineligible to play a heroic role.

Velikhov and Legasov had different areas of expertise—Velikhov in nuclear fusion, Legasov in noble gases—that had little to do with nuclear reactor accidents, but each had the inestimable advantage of having patrons on the Politburo. Velikhov was Gorbachev's scientific adviser, which, given the challenge of the American Strategic Defense Initiative, was a post of considerable political significance. Within the realm of science, he was an exponent of *glasnost* and *perestroika*, cooperating with the Americans on joint ventures in nuclear fusion. He was also vice president of the Academy of Sciences and the chief rival to Legasov to succeed Alexandrov.

Legasov was a protégé of Ligachev's and his father had been the head of the ideological department in the secretariat of the Central Committee, which Ligachev now controlled. Although aware, like Velikhov, that Soviet science was in need of some reorganization, Legasov was ideologically sympathetic to Ligachev's line, but behind his political differences with Velikhov, there was professional rivalry. Alexandrov could not go on for ever, and only one of them could succeed him as director of the Kurchatov Institute and president of the Academy of Sciences. Accompanying this professional rivalry, there was personal antipathy. Velikhov and Legasov disliked each other, and each tried to go to Chernobyl when the other was not there.

For Legasov, Chernobyl presented an opportunity to shine. Theoretically relieved by Velikhov when Silayev replaced Scherbina, he nevertheless remained to advise the commission. More commanding in demeanor than the pudgy Velikhov, he got greater respect from military commanders like General Ivanov of the civil defense and Major General Pikalov, commander of the chemical troops. He also had the advantage of being among the first to reach Chernobyl after the accident, so there was no questioning his undoubtedly heroic role. This record, together with his scientific reputation and political pedigree, made him the obvious choice to lead the Soviet delegation to Vienna. There was some opposition; hard liners in the Ministry of Medium Machine Building saw no reason for any disclosures at all and thought Legasov dangerously independent. But with the support of Ligachev, the Grand Inquisitor of the Central Committee, the Politburo approved him.

Like Ligachev, Legasov believed that there must be limits to *glasnost:* nothing should impair either the power or prestige of his country. There was also the question of his own responsibility for the catastrophe at

Chernobyl; as Alexandrov's first deputy, he shared the older man's responsibility for nuclear power, for he too had assured the world that Soviet reactors were entirely safe. At the time of the accident, he had been investigating the whole question of industrial safety, in particular the safety of nuclear power, which, since the Israeli bombing of the Iraqi reactor, had seemed vulnerable and dangerous in the event of war.

With respect to the report for the International Atomic Energy Agency in Vienna, Legasov's interests and those of his country were the same. Nor were any of his colleagues in any of the many institutes connected to the atomic-power industry likely to oppose him. It was his task to collate the reports that came in from the various institutes, and he worked tirelessly, the floor of his living room covered with piles of papers. It was a task of the utmost complexity, but he was helped by his colleagues in the Kurchatov Institute nearby.

That June, before the appointment of the working party, two meetings of nuclear physicists were held to consider ways in which the analysis contained in the report to the commission could be changed. The list of serious deficiencies in the reactor's design must be replaced with an explanation that put the blame on the operating personnel.

Given the nature of the Soviet system—the many tentacles of the octopus whose brain was the Politburo—the presentation to the public of an authorized version was not left to Legasov alone. In Kiev, Volodomyr Yavorivsky, part journalist, part publisher, part poet, and a secretary of the Writers' Union, set off for the thirty-kilometer zone to collect material for his novel, *The Star Called Wormwood,* in which the personnel at a nuclear power station were depicted as incompetent and dissolute. It was a roman à clef: the unfortunate people from Pripyat who had lost their jobs, their homes—some their health, a few their lives—were the models for thinly veiled portraits of idle operators, a drunken director and an adulterous chief engineer.

In Moscow, Legasov's friend Vladimir Gubarev, the scientific editor of *Pravda,* was suddenly inspired to write a play. Already, on 15 June, he had quoted in his paper the opinion of a Ukrainian party leader that "in the complex situation of the accident, former AES Director V. Brukhanov and Chief Engineer N. Fomin were unable to ensure the correct firm leadership and proper discipline, and showed irresponsibility and inefficiency." On 19 July Gubarev started to write his play, which he called *Sarcophagus*. He wrote it in a frenzy, sleeping only three hours a night, and finished it in a week. The drama was set in the medical wing of a radiological research center, where victims are brought in from a nuclear accident. A state investigator collects evidence from the patients. Little by little, the play exposes the carelessness and complacency of the plant's management, the recklessness of the operators and the cowardice of the personnel at the time of the accident.

The heroes of the play are a courageous young fireman and a young physicist who stays at his post. The villains are the director, who runs off to save his grandchildren but fails to warn the inhabitants of the town; the blustering, buck-passing officer responsible for safety in the plant; and the dosimetrist who, like the director, fails to warn others of the danger. The dramatic device of the state investigator enabled Gubarev to expose all the shortcomings in construction and management of the power station he had based on Chernobyl: the inflammable roof, the lack of adequate dosimeters and the delay in the evacuation of the town. As in Yavorivsky's novel, the personnel and the system are to blame, not the design of the reactor.

Although in essence it said little more than the article by Lubov Kovalevskaya published before the accident in *Literaturnaya Ukraina*, Gubarev's *Sarcophagus* broke new ground in speaking so openly about the shortcomings of the system. "What swine, excuse the unliterary word," says one of the characters, "switched off the emergency system? I wanted to say that this is murder. Not suicide, but murder . . . "

"The main thing for you," replies the physicist, "is to clarify who took off the emergency protection."

"Who took it off? Who took it off? It was the system that switched off the emergency cooling. A system of irresponsibility."

Gubarev was no dissident; the system criticized was not the Soviet one as such but that which had prevailed during Brezhnev's era of stagnation. Although larded with lofty thoughts about the nature of civilization and the destiny of man, Gubarev's play was no more than a plea for the new party line that Gorbachev was pursuing within the Central Committee. It was not communism but the lack of *glasnost* and *perestroika* that had led to the accident at Chernobyl.

As a result, *Sarcophagus* had no difficulty in finding theaters to stage it. It was noted at the time by William J. Eaton of the *Los Angeles Times* that "the play's rapid appearance suggests strongly that it has the support of high-level Kremlin officials, if not that of the Soviet leader, Mikhail S. Gorbachev himself." It appeared in the literary magazine *Znamya,* was translated into several foreign languages and appeared in many theaters abroad, establishing effectively in most people's minds that the accident was the fault of the operating personnel.

8

Others joined the campaign. In an interview in *Pravda* at the end of July, the head of the State Committee for the Use of Atomic Energy, Andranik Petrosyants, said that it appeared that the operators "had forgotten what kind of fuel they were working with."

But for the Soviet people more eloquent than any interview was the public disgrace of Brukhanov. For almost a month after the accident, he had remained at Chernobyl, fulfilling his duties as director of the station as best he could. He had been at the disposal of the government commission, and even when he suspected that the measures recommended by the illustrious academicians were futile, he carried out his orders to the letter. Shocked by the catastrophe and horrified by the deaths of his men, he was fully aware that as director he would sooner or later be held responsible for what had occurred.

On 22 May Brukhanov asked for permission from Mayorets, the minister of energy and electrification, to take a few days off to visit his wife and son, who had been evacuated to the Crimea. Permission was granted, and he stayed with them for a week. On 2 June he was summoned back to Kiev. There he was given a ticket to fly to Moscow the next day to attend a meeting of the Politburo in the Kremlin. When he went to say good-bye to the second party secretary, who had treated him with the utmost formality in the past, the man impetuously embraced him, which suggested to Brukhanov that something unpleasant had been arranged for him in the capital city.

The meeting started at 11.00 A.M. with Mikhail Gorbachev as chairman. First, Scherbina presented a report. Brukhanov was then invited to speak, and gave a brief explanation of what had occurred. Gorbachev asked him if he had known about the accident at Three Mile Island, and Brukhanov said that he had. He was impressed by the courtesy of the general secretary, noting that the higher the party official, the better he behaved. At the city level, they had snarled and shouted at him, at the district level they were a little more civil and here in the Kremlin, they were calm and polite.

The meeting continued until 7:00 P.M. At the end of it Gorbachev pronounced sentence: Brukhanov was dismissed from the party.

Brukhanov returned to Kiev with a foreboding that worse was to come. As a party member, he had been protected from arrest; this rule had been instituted by Khrushchev to prevent arbitrary action by the KGB. But now that he was no longer a party member, the legal process could begin. In Kiev, where a room had been reserved for him at the Leningrad Hotel, he was summoned each day to the office of the public prosecutor to draw up a statement, which when finished ran to ninety pages.

When it was completed, Brukhanov was driven back to the Pioneer camp, and it was only when he got there that he realized that he was no longer the director of the Chernobyl nuclear power station. No one told him, but he discovered that in his absence he had been replaced by Erik Pozdyshev. He still had some weeks' holiday to his credit, so he remained at the camp for the rest of June and throughout July with nothing much to do. Occasionally he would travel to Kiev to visit his wife, who was in the hospital for a heart complaint.

When it was time to return to work, Brukhanov asked if he could work as deputy chief of the industrial-technical unit, but his request was refused. On 12 August the deputy chief engineer, Smyshlyaev, told him that he was to report to room 203 at the prosecutor's office at 10:00 A.M. the next day. There he was charged under two articles of the Criminal Code of the Soviet Union. The first, Article 220, Section 2, covered the breach of safety regulations at "explosive-prone enterprises or . . . workshops"; the second, Article 165, Section 2, covered the abuse of power for personal advantage, and related not to the accident itself but to the signing of a document containing false data on the levels of radiation that followed it.

There was a break for lunch. When Brukhanov returned, two men had joined the prosecutor's team, and he was told that he was to be taken into custody by the KGB, where he would receive better treatment than if left in the hands of the militia. He complained that this was unnecessary because he was unlikely to abscond; only later did he realize that it was to prevent him from taking his own life. He was driven off in a jeep by the two officers from the KGB. When they reached the prison he gave up his belt, shoelaces and watch before being locked in a cell measuring two meters by four. There was a lavatory, a washbasin and, in the steel door, a spy hole, which opened every minute.

Brukhanov was to remain in this cell for a year awaiting trial.

On 21 July, before Brukhanov's arrest, the Politburo's full and final judgment was published in *Pravda*:

> "It was established that the accident occurred because of a whole series of crude violations of the rules for the operation of reactor installations that were perpetrated by the workers of this power station."

Experiments on the turbogenerators had been carried out without coordination or authorization from the appropriate organizations, and while they were being carried out, proper control was not exercised and appropriate measures were not taken. Both the Ministry of Energy and Electrification and the State Committee for Safety in the Atomic Power Industry were chastized for allowing such laxity at the Chernobyl nuclear power station, which had led to the death of twenty-eight people, the sickness of many more, the destruction of a reactor, the pollution of one thousand square kilometers of land, all to a value of around three billion rubles.

Promising to hand over the evidence collected by its own inquiry to the prosecutor's office, the Politburo then announced the sentence it had served on those it held responsible. Yevgeny Kulov, the chairman of the State Committee for Safety in the Atomic Power Industry; Gennady Shasharin, the deputy minister of energy and electrification; Alexander Meshkov, Slavsky's first deputy minister of the Ministry of Medium Machine Building and Professor Ivan Yemelyanov, Dollezhal's deputy director at NIKYET, were all dismissed from their posts. Anatoli Mayorets, the minister of energy and electrification, would have been dismissed if he had not been in his post for such a short period of time; as it was, he received a severe reprimand and was threatened with more serious punishment if he did not learn from his mistakes. Only Victor Brukhanov was expelled from the party. The fates of lesser figures, such as Dyatlov and Fomin, were left to the Central Committee of the Communist Party of the Ukraine. No disciplinary measures were taken against any scientists at the Kurchatov Institute.

9

This apparently impartial investigation by the Politburo, followed by the punishment of the culprits, cleared the decks for the Soviet team led by Legasov to present its case in Vienna. Few remarked upon the inherent contradiction in the Politburo's judgment: if the accident was caused by the breach of safety regulations by the operators, which in turn derived from the failings of Brukhanov, then why punish Dollezhal's deputy, Professor Yemelyanov, or Slavsky's deputy, Alexander Meshkov, at the Ministry of Medium Machine Building?

On 25 August the meeting of the world's experts on atomic energy was convened by the International Atomic Energy Agency in Vienna. In the four months since the accident, Chernobyl had aroused an obsessive and terrified curiosity throughout the world, so that as well as delegates from the accredited governments, thousands of correspondents descended on the ornate capital of the former Austrian empire.

It was sometimes a struggle to attend the deliberations, for there were only four hundred seats for delegates from 150 accredited governments. Legasov arrived with a large team, among them Guskova, Ilyn and Abagyan. The United States was limited to eighteen delegates, and so in order to attend, an independent expert on atomic safety, Professor Richard Wilson, first arranged to be included in the Kuwaiti delegation and then, when this was considered inappropriate by the State Department, was accredited as a correspondent of the British scientific magazine *Nature*.

The conference opened with Legasov's report; he spoke without faltering for over five hours. After a prologue expressing Gorbachev's good intentions, he proceeded to describe the RBMK reactor without mentioning any deficiencies in its design. He listed the different safety systems, suggesting that the comparatively slow rate at which the independent regulatory rods were inserted into the core when the emergency system was triggered "was offset by their large number." He described the tests made on the turbines in the early morning of 26 April, how the power surged out of control, leading the operators to press the emergency AZ button, but how "in the conditions that had now arisen, the violations committed by the staff had seriously reduced the effectiveness of the emergency protection system. The overall positive reactivity appearing in the core began to increase. Within three seconds the power rose above 530 megawatts, and the total period of the excursion was much less than twenty seconds. The positive void coefficient worsened the situation."

In analyzing the causes of the accident, Legasov insisted that "the design of the reactor facility provided for this type of accident, with allowance for the physical characteristics of the reactor, including a positive steam void coefficient of reactivity." He then listed "the most dangerous violations of the operating rules committed by the staff of the fourth unit of the Chernobyl nuclear power plant: . . . The chief motive of the staff was to complete the tests as expeditiously as possible. The failure to adhere to instructions in preparing for and carrying out the tests, the noncompliance with the testing program itself and the carelessness in handling the reactor facility are evidence that the staff was insufficiently familiar with the special features of the technological processes in a nuclear reactor and also that they had lost any feeling for the hazards involved."

The designers had not provided a protective safety system to prevent an accident of this kind because they never envisaged "the deliberate switching off of technical protection systems coupled with violations of the operating regulations"; they had "considered such a conjunction of events to be impossible. Thus, the prime cause of the accident was an extremely improbable combination of violations of instructions and operating rules committed by the staff of the unit."

Next, Legasov described the measures taken to protect the population from the effects of the accident, and announced that 135,000 people had been evacuated from the Pripyat and the surrounding zone. . . . "As a result of these and other measures, it proved possible to keep population exposures within the established limits. The radiological effects on the population in the next few decades were evaluated. The effects will be insignificant against the natural background of cancerous and genetic diseases."

It was a virtuoso performance and was greeted with great acclaim. Together with the charts, diagrams and statistics provided by the Soviets, the impression was given that the full and frank disclosure promised by Gorbachev had been delivered. "At last," said the headline in a Viennese newspaper, "a Soviet scientist who tells the truth." Not only did Legasov seem to speak with much greater candor than the world had come to expect from behind the Iron Curtain, but there were moments when emotion showed through the traditionally impersonal manner of the Soviet spokesman—for example, when Legasov said that the commission would like to have heard more from one of the operators (Akimov), "but unfortunately he died."

The next day it was Guskova's turn to describe the treatment that had been given to her patients in Hospital No. 6. Some of the cases made medical history: no one had imagined that someone as severely irradiated as Piotr Palamarchuk or Anatoli Tormozin could survive. Robert Gale was with the United States delegation when Guskova stated that this was not due to bone-marrow transplants, which both men had rejected. He felt that political pressure had been put on Guskova to downgrade American assistance.

Given the publicity he had received in his own country, Gale had become something of a hero. He seemed to symbolize the superiority of American medicine, and hence the American way of life. Therefore it was galling to hear from Guskova that while the Soviets were grateful for his help, and particularly for the equipment donated by Armand Hammer, it had not actually saved a single life.

Guskova, on the other hand, who had been annoyed by the way Gale, an expert on bone-marrow transplants, had been presented by the media as a specialist on radiation sickness, noted with some satisfaction that her fellow specialists on the American delegation seemed to share her irritation with the man.

. . .

After Guskova, there was an address by the jovial Armenian, Professor Armen Abagyan. His institute, albeit under the umbrella of the Ministry of Medium Machine Building, was nevertheless distinct from either Dollezhal's bureau or the Kurchatov Institute. Specifically set up after the accident at Three Mile Island to supervise safety in the atomic industry, it had recognized the deficiencies in the RBMK's design, but given the climate at the time, did not insist on its analysis of the accident appearing in Legasov's report.

Indeed, Abagyan, when questioned by experts from Canada about the emergency-shutdown systems operating at Chernobyl, avoided answering; to the anger of many of the delegates, he appeared to pretend that he did not understand. When asked what measures would be taken against those responsible for the accident, he replied that punishment did not fall within his area of expertise. It was as close as he could come to suggesting that the operators alone were not to blame.

Wednesday, 27 August, was an open day during which experts from different countries could question their Soviet colleagues. Professor Wilson approached two members of the Soviet delegation—a professor from the Institute of Applied Geophysics and a young researcher from the Kurchatov—and asked if he could compare his own data on the release of radioactivity with theirs. There followed an open and straightforward discussion that lasted for three hours. They described their methodology to Wilson; it broadly coincided with his, but when he checked the total recorded emission with the sum of figures on the printout, he found that they did not match.

The two Soviets looked confused and told Wilson that there must have been some mistake in the calculation. In reality, six pages covering the serious contamination of the area north of Gomel had been removed. Wilson was later told that this had been done on Legasov's instructions.

IX

1

While the political damage done by Chernobyl was contained at Vienna, work continued on the site to clear up the mess. On 11 May Ivan Silayev, who had succeeded Boris Scherbina as chairman of the government's Chernobyl commission, broadcast to the nation that "the main threat from the destroyed Chernobyl nuclear power station has gone." Academician Velikhov, also without spelling out quite how terrible things might have been, confirmed that "we really do see today that we are not threatened by the major catastrophe we feared." Now they would move on to the next stage, "which consists of total—well, putting it crudely—deactivation of the territory, and the conservation and encapsulation of all the radiation."

After the blanket of secrecy that had covered Chernobyl in the weeks following the accident, a policy of relative *glasnost* now prevailed. On 17 May on Moscow television Silayev described in some detail the building of the heat exchanger under the damaged reactor, a "concrete cushion," and announced the construction of a structure to entomb the radioactive core. He called it "a sarcophagus . . . a huge container, let us say, which will enable us to secure the burial of everything that remains of the radioactive fallout of this entire accident in general, so that it can be reliably contained and supervised and will not give rise to any doubts whatsoever."

To design this sarcophagus, Silayev enlisted a team from Leningrad's All-Union Research, Design and Development Institute of Power Engineering. It was a complex task, not simply because the sarcophagus had to be built in such hazardous conditions but because the corpse to be buried was still twitching; plutonium 239 has a half-life of 24,360 years. The graphite had burned out, and the temperature in the core had declined to about 270°C, but the fuel was still there in an unknown condition. It required no oxygen for fission to resume; therefore it could not simply be buried. The sarcophagus would have to contain the radiation, yet have apertures for ventilation and observation.

Nor was the core itself the only source of highly dangerous radioactivity. Lumps of graphite and fragments of the fuel itself had been spewed out by the explosion. These were difficult to find, and it was highly dangerous to walk around the site waiting for the needles on the dosimeters to jump to a level of two thousand rads per hour.

General Pikalov, commander of the chemical troops, developed a better method. The day after he arrived at Chernobyl, his men had taken aerial photographs of the power station, which were then collated to form a photographic map of the whole area. On this Pikalov noticed small glowing spots, which were discovered to be fragments of nuclear fuel. These were covered with rubble, concrete and concrete slabs by the arm of the tracked EMR2 reconnaissance vehicles, and by eighteen ton remote-controlled bulldozers that had been flown in from Chelyabinsk in cavernous Ilyushin 76 transport aircraft.

More of a problem were the lumps of highly radioactive graphite that lay on the roof of the third unit. The Soviet Union did not possess the robotic equipment to throw them over into the crater of the ruined reactor, so small mobile robots were purchased from Germany and flown to Chernobyl. They did not work. The bitumen roof was too soft; the robots either got stuck on the mushy surface or were stopped by the blocks of graphite themselves.

There was only one solution: the task had to be done by humans. The dosimetrists made the calculations. Wearing protective clothing, one chemical trooper could spend between one and two minutes on the roof before accumulating a dose of twenty-five rems. Pikalov, who had faced German bullets at the age of fifteen, saw no reason to doubt a comparable heroism in the young men under his command. The enemy might be invisible, but the call of their Soviet motherland was the same.

It was done. Television cameras were set up so that the officers could show their men where they should go and what they should do. The soldiers wore special lead plates to protect their testicles, hands, feet, chests and backs; goggles and respirators covered their faces. Like contestants in a grotesque race, the young soldiers in their costumes weighing 55 lbs climbed the staircase leading up to the roof; then, one after the other, each man ran out with a shovel, scooped up the graphite or fuel-rod fragments, rushed to the edge, threw the lumps over into the abyss and ran back to take cover behind the concrete wall. "Whoever you speak to," Silayev reported to the cameras of Moscow television, "they say, 'We understand what the homeland requires of us and we will do it.' Therein lies the heroism of our Soviet man."

At times Soviet man required reassurance; with as many as 260,000 men working over the months as liquidators in the zone, not all showed the same degree of courage or insouciance. Pikalov's troops became afraid that the food they were given was contaminated, and the general had to go to their camp and eat from the common pot. Detachments were summoned from all the Soviet republics, and some were reluctant to go—particularly those from the Baltic states. In Latvia wives tried to prevent their husbands from leaving home by locking the doors of their flats. Members of the Estonian brigade were so dissatisfied with their encampment, and the absence of dosimetric data to show the risk they were taking, that they rioted and marched to the railway station, ignoring their officers' commands. Subsequently there were rumors—broadcast by the Voice of America but emphatically denied by the Soviets—that the twelve leaders of this demonstration had been shot.

To improve the morale of the Latvians, the local leadership asked the Writers' Union for three volunteers to bring books to the bored reservists in the forbidden zone. There was no rush to come forward. A writer in Riga, Marina Kostenecka, was waiting for a bus when a colleague asked if she would like to go to Chernobyl in a couple of days. Curiosity overcame prudence; she not only went to the Latvian camp, which was thirty kilometers from the nuclear power station, but persuaded the commanding officer to take her to the power station itself. On the way he showed her a birch tree on which an army medical officer had supposedly hanged himself in despair.

Marina saw the station and, by moonlight, the abandoned city of Pripyat. Surrounded by an electrified fence to deter looters, it was eerily silent. As she stood considering that this was how the whole world would look after a nuclear war, a cat came through the fence, mewing at the sight of a human being. When she leaned over to stroke it, the colonel grabbed her arm; the cat's fur would be highly radioactive.

2

As the liquidation proceeded, some of the soldiers fell ill. None suffered from radiation sickness as such, but by early June Academician Ilyn was obliged to concede that some had been treated for lesser ailments, such as nausea and fatigue. The medical commission revised its recommendations for a maximum permissable dose from twenty-five rems to ten, and for a lifetime dose from fifty rems to thirty-five. Fearing for the genetic consequences, young conscripts, whenever possible, were replaced by older reservists, known in army jargon as "partisans." The danger was not just from the graphite and fuel that had been thrown out of the reactor by the explosion but from the highly radioactive dust that had fallen throughout the zone. Time and again, the liquidators had to hose down the houses in Chernobyl, Ivankov and the Pioneer camp. Some vehicles were irretrievably contaminated and had to be dumped in huge pits—not only buses and trucks, but hundreds of cars from Pripyat, bought after many years of saving and waiting. Other vehicles were decontaminated; the huge black Zils that had brought Ryzhkov and Ligachev to Chernobyl were used to ferry people to and fro within the zone.

Contaminated topsoil had to be removed by bulldozers and buried. There was also a large swath of forest that had been singed by the plume of radionuclides; the green conifers were now brown, yellow and red. They were felled but could not be burned, because the smoke and ash would themselves be radioactive. They were buried in vast pits that had to be lined with concrete to prevent the radionuclides from seeping into the soil.

More immediate was the possibility that the radionuclides on the surface would be washed down into the streams and rivulets that fed the Pripyat and the Dnieper. The Kiev Sea, only a few kilometers from the power station, supplied that city with drinking water, and the waters of the Dnieper itself, on its course to the Black Sea, were used by fifty million people.

As a precaution, in Kiev boreholes were sunk to the water table in the city's main bakeries to ensure uncontaminated water for the baking of bread. At Chernobyl, the military engineers embarked upon two major projects: the first to build a new bridge over the Pripyat, the second to dam the river itself. Time was short; there was a forecast of rain in ten days. The most optimistic estimate was that it would take two months to dam the river, but by a concentration of all the country's resources, it was done in little more than a week.

The new bridge was built with comparable alacrity, as was a vast reservoir to store radioactive groundwater caught by a number of smaller dams and channeled through a network of drains. A subterranean wall was built between the reactor and the river to contain the contaminated groundwater, and three hundred boreholes were sunk so that if the water table should become contaminated it could be pumped dry. Four hundred men worked on this, coming in from camps outside the thirty-kilometer zone. They worked shifts of twelve hours, frequently changed their clothes—particularly their boots—and wore film-badge dosimeters which they could not read themselves, but which were checked by technicians when they returned to their quarters each day.

The war against radiation was waged in the air as well as on the ground. The high level of contamination northeast of Gomel had shown how disastrous it could be when rain brought the radionuclides back to earth. Once, when rain clouds approached Kiev, planes from Professor Israel's Committee of Hydrometeorology fired pellets in an attempt to precipitate rainfall before it reached the city, but the clouds were blown in a different direction and the downpour predicted never occurred.

There was a constant movement of helicopters throughout the zone: Pikalov's taking measurements within the ten-kilometer zone, Israel's doing the same in the thirty-kilometer zone and beyond, Antoshkin's spraying the ground with a plastic solution that subsequently solidified, thus trapping the radioactive dust. For the last purpose, huge transport helicopters were fitted with tanks holding twelve thousand liters of the polymer solution. There were still constant flights over the reactor as scientists took readings to try to learn what was going on in the core.

Professor Ryzantzev from the Kurchatov Institute sometimes made two or three flights a day. He noticed the anxiety of the young pilots who, though their helicopters were now fitted with sheets of lead to protect them from radiation, were afraid that it might make them unable to have children. After a run over the reactor, seeing one young pilot peer up to the light to try and read the gauge on the pencil-like dosimeters, Ryzantzev asked, "Are you a pilot or a sailor?"

"A pilot. Why?"

"Then why are you looking through a telescope?"

"I'm worried about the dose."

"You don't need to worry. I've been working with reactors for more than twenty years, and look at me. As long as you don't exceed a dose of twenty-five rems, you'll be perfectly all right."

"In that case," said the pilot, "can we make another run?"

"Why? We've just returned."

"Because if I pick up another three rems, I can go home."

To avoid making these repeated sorties, Ryzantzev and his colleagues in the Kurchatov Institute designed a thermocouple to monitor the reactor. Nicknamed the "needle," it was ten centimeters in diameter, eighteen meters long and painted with black and white stripes. Filled with instruments of different kinds, it was to be lowered twelve meters into the reactor core and the data from the instruments trasmitted data through a three-hundred-meter cable—the "thread." Despite the enthusiasm of the young air force pilots, their Mi-8 helicopters could not provide the stability that Ryzantzev required to lower this needle through the crust of sand and boron into the small aperture left by the dislodged biological shield. Therefore Silayev, at one time the minister of aviation, summoned a unique helicopter, the Ka-27, with double rotating blades, which had been developed by the ministry's experimental design bureau. This could hover with great stability for long periods of time. The chief of the institute, Igor Erlich, arrived with his deputy, Eduard Korotkov, and three of his ablest test pilots, among them Nikolai Melnik.

After several trial runs over a mock-up of the reactor, Melnik took off with the needle trailing from the cable attached to the Ka-27. At the first two attempts, the needle failed to penetrate the crust of sand, boron and lead. The third time, he was lucky; it sank to two-thirds of its length. The three-hundred-meter cable was released and fell to the ground. However, it fell on the wrong side of the reactor, on highly contaminated ground. A team of dosimetrists approached it in one of the EMR2 reconnaissance vehicles, but the cable was caught on the roof, out of the reach of its mechanical arm. Therefore they returned to the fourth unit and climbed up to a corridor where the dangling cable could be seen from a plate-glass window. They smashed the window, grabbed the end of the cable and dragged it into the corridor, where it could be connected to the network that Rzyantzev had prepared.

The data received from the needle confirmed what was already suspected—that there was no further danger from the fuel that remained in the reactor. Already, by the end of June, the tunnel had been cut through the sandstone beneath its foundations and the heat exchanger—a network of water channels set in a huge concrete "cushion"—completed. Academician Velikhov had returned to Moscow, and the miners and metro workers had gone home. With the temperature inside the core now down to 150°C and the original base to the reactor intact, the heat exchanger remained as a monument to the real danger of the "China syndrome" during those dramatic days in early May.

Aboveground, the construction of the sarcophagus proceeded, but with considerable difficulty because of the extremely high levels of gamma radiation emanating from the reactor. To be able to approach the site, the scientists decided to build a thick concrete wall. A factory in Kiev constructed large steel tanks measuring eighteen by six by six meters, which were placed on trailers and towed into position in front of the reactor by the EMR2 reconnaissance vehicles. Soldiers then shot out the tires with machine guns and the trailers sank to the ground.

A relay of trucks brought ready-mixed concrete to the site, and different methods were tried to fill the metal boxes, but all proved unsatisfactory. Cranes with buckets were too slow. They tried pumping it in, but to do this the pebbles in the gravel could not be more than thirty millimeters in diameter, and this made the mix so liquid that it oozed out of the sides of the containers. Next they tried a system of conveyor belts, but the vibration separated the gravel from the cement.

In desperation, Mayorets, the minister of energy and electrification, turned to his own All-Union Research and Development Institute for New Power Enterprises and its Department of Accelerated Methods of Hydropower Construction, a team that developed new techniques in the construction of dams and power stations. This was led by an engineer named Robert Tilles, who had acquired considerable experience in the use of concrete when building hydroelectric power stations in Siberia.

A small, quiet man with a strong character, Tilles was in the hospital suffering from blood poisoning when Mayorets summoned him in the middle of May. There was a conference at his bedside, and the next day Tilles, his chief assistant and one of their specialists, Igor Kravchenko, flew to Chernobyl in the minister's plane.

The solution Tilles proposed was to bring in equipment that had recently been acquired for $3 million from the Rotec Corporation in the United States: vibration-free conveyor belts and concrete layers. They had first used them in the construction of the Ingouri hydroelectric power station in the Georgian Republic. Galling though it was for the government commission to have to resort to American equipment, it was dispatched from a depot in Lithuania while a team flew in from the Caucasus under the direction of Igor Kravchenko and another of Tilles' experts, Dimitry Shatalov.

Because of the hazardous nature of the work, some of those whom Tilles invited refused to go to Chernobyl. There were also delays in supplying material. Some of the drivers delivering the ready-mixed concrete were so frightened that they drove off the road, dumped their load, and returned to their depot in Kiev. (Later the situation improved when a plant was built between Chernobyl and the Pripyat River.) Guided from a lead-lined cabin on a trailer near the site, the Rotec conveyors delivered the concrete into the sixty-ton containers, which like giant bricks built a "biological protection wall." By the end of July it was completed, and work on the sarcophagus itself could begin.

3

The containment of the consequences of the accident was directed by the government commission. The chairmen, all deputy prime ministers, served a limited tour of duty in Chernobyl itself; Boris Scherbina was replaced by Ivan Silayev, Silayev by Lev Voronin, Voronin by Yuri Maslyukov; Maslyukov by G. Vederniko. Everyone was under the control of the commission, but beneath them during this critical period a camaraderie arose among those of different rank. All suffered in the same way from the radiation, with running noses and squeaky voices. They were bound together by their common purpose. Marshals and ministers deferred to the experts; a colonel might be found driving an EMR2 comandeered by an engineer.

The eminent academicians, Legasov and Velikhov, and beneath them the scientists from the Kurchatov Institute and NIKYET exercised the authority that came from their expertise. At a more practical level there was rapport between the army and the operators of the Chernobyl nuclear power station. Each group relied upon the other. At an early stage, when a possible alternative to building the heat exchanger under the reactor was simply to pump concrete into the empty bubbler pool, an order was relayed to Nikolai Steinberg and a former submarine officer, Reichtman, to survey the approaches to the bubbler pool in the bowels of the reactor and find the best place to blast a way in.

The levels of radiation were between fifteen and twenty rems an hour, but in some places—for example, by one elevator shaft—they rose to two hundred rems per hour. Steinberg had his own intuitive dosimeter: where the level was over 135 rems per hour, he could feel it like a punch in the eye. He also knew how to dodge from behind the shelter of a pillar to that of a concrete wall. It was stiflingly hot in their respirators and rubber suits, and they were soon up to their ankles in sweat.

When they returned to the bunker, they found a group of soldiers led by a small man with a strong, determined face. He barked out an order to Reichtman and Steinberg: the detonation to blow a hole through the wall was to take place at precisely four that afternoon, and the operation would be reported to the Politburo.

"And who are you?" asked Steinberg.

The officer's eyes widened at the audacity of the question. "I am Marshal Oganov, commander of the military engineers!"

"Very well," said Steinberg. "I will show your men where to lay their charges."

Three holes were blown in the wall of the fourth reactor, but the way through to the bubbler pool was still blocked by pipes and machinery. A further explosion was thought too risky: it might raise a new cloud of radioactive dust. Therefore Kizima, the head of construction at Chernobyl, proposed cutting a path through the obstruction with a welder's arc. Cables were laid to supply electricity so that Kizima's men could begin the next morning.

At 4:00 A.M. Vadim Grishenka saw smoke coming from the ruins of the fourth unit. The alarm was given. Fire engines came from Ivankov, and Nikolai Steinberg was summoned from his bed in the Pioneer camp. At 5:00 A.M. a helicopter flew over the unit to try to pinpoint the source of the fire, but all the pilot could see was smoke. Therefore a group of three operators—among them Grishenka and Steinberg—went through from the third unit. They found that short circuits had set fire to the electric cables, so they disconnected them and left the cables to burn themselves out.

There had to be a scapegoat for this second fire, and the man chosen was Taras Plochy. The accident at Balakovsky was still held against him; therefore he was sent back to resume his duties there as a deputy chief engineer, while the position of chief engineer at Chernobyl was given to his friend and erstwhile protégé, Nikolai Steinberg.

A number of hazardous sorties into highly contaminated areas were undertaken by teams of operators and soldiers. No one could be sure that Lelechenko had succeeded in cutting off the supplies of hydrogen at the time of the accident, since he had not entered his mission in the log. A manned EMR2 was sent to cut the pipes and let the hydrogen seep into the ground. Vladimir Nesterov, the deputy head of the workshop, went back into the turbine hall of the fourth unit, where the radiation was eight hundred rems per hour, to check that the oil tanks were empty.

Toward the end of May, the KGB decided that to learn the cause of the accident it must have the Mercedes-Benz equipment that Metlenko, the engineer from Donenergo, had brought to Chernobyl for the tests on the turbines. The equipment was retrieved with two EMR2 vehicles, one driven by Pikalov himself, the other by the deputy chairman of the KGB, General Sherbak. Because their short-wave radios would not function, the two communicated with each other via Moscow on special frequencies, and managed to drag out the trailer with the precious equipment.

Operators helped the chemical troops locate the fuel that had been spewed out of the reactor and helped design a device to pick up the smaller pieces of radioactive graphite—a greased wire mesh stretched across a metal box, which was lowered by the eight-meter arm of an EMR2. While engaged in this operation, a lieutenant of the chemical troops came to a spot a hundred meters or so from the reactor where the level of radiation suddenly leaped from eighty rems to 1850. He returned to the bunker, his face pale. Steinberg and the other engineers studied the photomap made by Pikalov's chemical troops and realized that he had come across a fragment of fuel. This was reported to the commission in Ivankov.

At eight the next morning, three officers arrived at the bunker—a colonel, a major and a naval captain—and reported to Steinberg, the chief engineer. "We have been ordered to remove the fuel. Show us where it is."

"Do you know what you are dealing with?" asked Steinberg.

"Our orders are clear. If we haven't removed the fuel by six this evening, we are to be demoted."

"That might be better than dying," said Steinberg.

"Please show us where it is."

"Who issued these orders?"

"General Malkevich."

"Who can change these orders?"

"Only General Gerasimov, chief of staff of the Southwest area."

Steinberg telephoned Gerasimov and suggested that the order be rescinded. Gerasimov agreed, and the three officers returned to their base.

In due course, the fuel was covered, first with iron ore and then with concrete without risking anyone's life. The would-be heroes whom Steinberg had relieved were grateful for what he had done. Four hours after they left, a colonel appeared in the bunker and asked for Steinberg.

"Are you Steinberg?"

"Yes."

"I have a package from headquarters."

Steinberg opened the parcel and found two bottles of cabernet.

4

While the operators assisted the military in the work of decontamination, their chief task was to bring the first three units back on line before winter, when the three-thousand-megawatt capacity would be urgently needed. The electrical circuits had to be repaired and the cooling systems overhauled, all in hazardous areas of the power station. Radioactive dust had entered all three units through the ventilation system. In the third unit, both of the electrical circuits in the cooling system were damaged; unpurified water had been used to cool the core during the shutdown because all the purified water been diverted to the fourth reactor.

The work was hazardous and exhausting. Many operators succumbed not so much to radiation sickness as to stress. Although they knew how to minimize their exposure to radiation by shielding their bodies from its source, they were also aware that it could not be avoided altogether. Both inside and outside the power station, there was constant monitoring of the level of radiation. At the Pioneer camp, readings were taken every hour, both at ground level and at one meter above it. Anyone who wanted to could read the statistics, but they were not published in the press. Indeed, the young doctor from Pripyat, Tatiana Ben, having seen that the level at the entrance to the nuclear power station was one rem per hour, later saw it stated in a newspaper as five millirems per hour. It came as a rude shock; hitherto she had believed what she had read in the press. She was now without her husband, Anatoli, whose hand had become swollen and sore after removing his gloves to treat the radioactive body of Sashenok in Pripyat hospital on the night of the disaster.

The operators all wore dosimeters, but few added up their cumulative dose. A blend of professionalism and patriotism drove them on, not just to eliminate the consequences of the accident but to get the first three units back into production. Whatever the risk, they stayed at their posts. In particular, Dyatlov's protégés from Komsomolsk worked tirelessly to repair the damage done unwittingly by their patron and friend. Sitnikov was dead, but the boyish Vadim Grishenka was now head of the workshop for the first and second units, and Vladimir Chugunov, who had almost died from radiation sickness at Hospital No. 6, had persuaded the doctors on his release to allow him to return to work at Chernobyl.

On 11 May a call came through to the bunker with the news that Akimov had died. Steinberg, who on leaving Chernobyl in 1984 had handed over his flat to the Akimovs, was so upset that he commandeered a jeep and drove to Pripyat. The traffic lights were working and music still came from the loudspeakers on the streets. Lines of washing flapped on the balconies; in a playground, by the swing and jungle gym, sat an abandoned pram. In the amusement park, the Ferris wheel stood idle.

It had been Brukhanov's dream to have a city filled with roses; now they were in full bloom. Steinberg stopped by Akimov's flat. He saw that the windows were open and realized that Akimov would never return to close them. At the entrance was an old pair of slippers and a badminton racket, both dropped in the rush of the evacuation. He turned the jeep around, returned to the power station and never went back to Pripyat again.

There were moments of comedy. When Steinberg attended his first meeting of the commission, he was amused by all the generals reporting in squeaky voices, but a day or two later his was the same. There was also a sense of camaraderie among the liquidators that eliminated all the usual prejudice and formality of everyday Soviet life. In this atmosphere of crisis, it did not matter whether a man was a Russian, a Tatar or a Jew, an officer in the army or the KGB, a party member or, like Steinberg's friend Reichtman, one who had been thrown out of the party for insubordination.

To compensate them for their hazardous work, every effort was made to ameliorate their living conditions. When Ryzantzev and his companions arrived at Ivankov, they were astonished to find caviar on the menu for supper. The soldiers lived under canvas, sixty men to a tent, and at the Pioneer camp the operators and scientists slept up to eight in a room. There was segregation according to sex, so married couples, like Anatoli and Tanya Ben, could not share the same room.

In July the housing shortage was eased by the arrival of eight comfortable cruise liners, which were moored fifty kilometers from the power station at a point known as Green Cape on the shore of the Kiev Sea. Built to accommodate Western tourists in some comfort, they had berths for up to twelve hundred workers from the Chernobyl power station. There were shops, libraries and a television on each deck. To mitigate the effect of radiation, a shift system was arranged for the personnel; after fifteen days at the power station, they were given fifteen days leave in Kiev. Wives could not join their husbands unless, like Tatiana Ben, they worked inside the zone. Whenever possible, particularly if they had children, women were evacuated to uncontaminated areas. However, there were some among the medical personnel, and others who still worked in ancillary positions at the plant. There were some love affairs among the liquidators, as well as transitory liaisons of a less romantic kind. When Luba Kovalevskaya returned to the zone during the summer to write about the living conditions of the power station personnel, she was appalled by what she saw. Dormitories were used as brothels; some had the vomit of drunks on the floor. She also heard stories of rapes, broken marriages, alcoholism and suicide.

Some bonds, however, were more elevated and had endured for many years. Katya Litovsky, the young woman who had worked for Steinberg in the turbine hall in the late 1970s, had been evacuated from Pripyat to her mother's cottage in the southern Ukraine. When she heard of his appointment as chief engineer announced on television, she leaped up and shouted, "Mama, Mama, now everything will be all right." To her mother's consternation, she immediately wrote to Steinberg volunteering to return to work at Chernobyl in any capacity whatsoever. Her offer was accepted, and though she was qualified as a mechanic, she took up the duties of secretary to the chief engineer.

Katya had been married and had two children, who went first to live with her former husband in Krasnoyarsk, then to a boarding school in Kiev. To her, working for Steinberg at the nuclear power station came before both family life and her own well-being. The conditions were hard; she wore a mask to work, and the windows of her office were covered with lead. Nevertheless, she was happy, and whenever she went away on leave she longed to return to Chernobyl. As one of the few women at the plant, she sensed her value to the workers' morale. "We were all a little in love with you," a naval officer told her later, "and tried to live up to the cheerful look on your face."

To raise their spirits, the operators also played tricks on the management, which would have incurred severe censure under normal conditions. Steinberg had a pass printed that certified that he had received "an external dose of radiation of forty-five rems and an internal dose of 40 percent alcohol. This cannot be explained. He is allowed to go nowhere. He is to be buried in the Chernobyl area at a depth of not less than eight meters. Drilling is compulsory. A biological protective wall must measure 4 × 4 × 0.5 meters."

Steinberg also gave himself certain rights: "1. To buy alcohol and have tests without standing on line. 2. To be in public places while dead drunk. 3. To attend women's beaches and saunas (sexually harmless). 4. To attach to his T-shirt any medals, orders or other objects that glitter. 5. To consume energy from the fourth unit free of charge. 6. He is immune from arrest by the administration; he is not obliged to dry out when drunk."

The order was signed by Bang, the vice chairman, and countersigned by the secretary, Mille Roentgen.

Steinberg was repaid in the same currency, receiving a report from the head of the scientific research unit that complained "that the female workers in the canteen, in the period from 30 August 1986 to 24 September 1986, have systematically contributed to the deterioration of the health of the army's scientific group No. 19772. As a result, the effectiveness of their work has sharply decreased, and therefore the deadline for the relaunch of the first two units has had to be postponed.

"I would therefore like you to instruct the head of the canteen to make sure that the female personnel change only in rooms that have no curtains and that have windows facing the office of the above-mentioned scientific group, and that you permit regular meetings between representatives of the canteen and our workers in the laboratory of the chemical workshop between 9:00 A.M. and 6:00 P.M. (telephone No. 2448).

"Scientific research has shown that if the first request is met, then the coefficient of labor productivity will rise in the above-mentioned military unit No. 19772 by 1.58 percent; if the second is met, it will rise by 82.52 percent; and if the above-mentioned is taken into account, the launch of the unit No. 3 might be brought forward by 14+ days."

The report was signed by "V. P. Karpov, Head of Scientific Research Unit."

5

Although brave about the risks they ran, the operators nonetheless took all necessary precautions to minimize their dose, and regulations in the zone were strictly applied. It was enclosed by a barbed-wire fence and patrolled by the MVD, the troops of the Ministry of Internal Affairs. There were only three points of entry, each of which was equipped with a medical clinic, dosimetric control and decontamination facilities. No vehicles that were "dirty"—that is, contaminated by radiation—were permitted to leave the zone. No one could enter without a pass, and everyone was encouraged to wear a mask or respirator to avoid inhaling contaminated dust.

Pripyat posed a particular problem. Almost all the possessions of the power station's relatively affluent work force remained in the several thousand apartments. The levels of radiation were too high for the militia to mount a guard on each block; therefore an electrified fence was erected around the town, and alarm systems installed in some of the shops and offices. This did not prevent some looting, and in due course it was decided to allow the inhabitants to return and salvage those belongings that were not badly contaminated. In this way Luba Akimov was able to retrieve her husband's car, which had been kept in a garage; she gave it to his brother, who had donated his bone marrow in a fruitless attempt to save Alexander's life.

The rural population also came back to retrieve their belongings, and in two villages where the contamination was low, Cheremoshnaya and Nivetskoye, residents were allowed to move back to their homes. Some had never left; in the village of Opachichi, fifteen kilometers from the reactor, Ivan and Irina Avramenko hid from the militia, and whenever they heard a helicopter ran for cover under a tree. They were both over seventy years old, and no talk about radiation was going to persuade them to leave the land that their forefathers had farmed for generations. In the town of Chernobyl, too, some of the older inhabitants sneaked back into their wooden houses. To prevent the liquidators from moving in, they put up little notices saying THE OWNER LIVES HERE, and the authorities had not the heart to move them.

Farther afield, forty kilometers to the east of the power station, work had started on a new town for the displaced power station workers, called Slavutich. It was built on virgin land near the east bank of the Dnieper between Chernobyl and Chernigov, and each of the republics that made up the Soviet Union was to design and build a section of the city. It was to be as bright and modern as Pripyat, with small villas for senior personnel. Though the homes were lavish by Soviet standards, the families for which they were intended were in no hurry to move in. Some of the wives in Kiev were unhappy at their husbands' prolonged absence, but having been allocated flats either in Kiev or Moscow, they began to dread the moment when they would be asked to return.

6

Like a modern cathedral, the sarcophagus rose to cover the ruins of the reactor. The body of the unfortunate Khodemchuk could not be recovered; never, since the time of the pharaohs, did one man have such a costly mausoleum. Tilles' wall at the base was eight meters thick: the concrete struts at the side were fifty-five meters high. Huge steel girders stretched from this new wall to the still solid structure of the new unit and were then covered in concrete to seal the roof. No one was on the site to guide the girders into place; it had to be done from the relative safety of the crane's cab, using television cameras to monitor the work. Ventilators were installed to cool the inside of the sarcophagus and were fitted with monitoring equipment and powerful filters.

On 23 September Boris Scherbina announced that the sarcophagus was virtually completed. "There are now no dangerous discharges from the damaged reactor. This enables us to start up the power station again." On 29 September the first unit received its certification, and on 1 October it went on line. The second unit followed on 5 November. Two thousand megawatts had been restored to the grid. There was to be a delay of more than a year before the relaunch of the third unit, but by the end of 1986 the various government commissions—Ryzhkov's in the Politburo, Scherbina's at Chernobyl and Shepin's in the Ministry of Health—could be reasonably pleased with what they had achieved. Certainly there had been a serious accident; one might even call it a disaster. Thirty-one men and women had been killed and many others injured or made ill. The cost was enormous; no one yet knew how many billions of rubles would be required to pay for clearing up after the accident and relocating people, let alone the loss of the fourth unit at Chernobyl and of six months' electricity from the other three. Nor could the cost be counted only in lives or in rubles: there was also the humiliation of the Soviet Union in the eyes of the gloating bourgeois capitalist world.

Yet even in this area there was reason for some satisfaction. Legasov's performance at Vienna had gone down well. The horrific exaggerations in the American press had been exposed as false, and the danger that public opinion would force them to close all their RBMK reactors, with catastrophic consequences for the Soviet economy, had been averted. In some quarters, indeed, the efficiency with which the city of Pripyat had been evacuated and the 135,000 inhabitants of the thirty-kilometer zone relocated was proof that there were some advantages to a political system with centralized power. Think of the accident at Three Mile Island, and consider the chaos that would have ensued if a comparable catastrophe had happened in the United States!

Only one thing remained to be done before the file could be closed: the culprits had to be punished. There had to be a trial. It was not a question of deciding on anyone's guilt or innocence—this had already been done by the Central Committee—but in the era of *glasnost* and *perestroika* there had to be a convincing semblance of legality. Justice must be seen to have been done.

An investigation had been opened on the very day of the accident by the Ukrainian public prosecutor in Kiev, and the next day by his All-Union opposite number in Moscow. The general prosecutor in Moscow, Alexander Reykunkov, appointed a senior assistant to head the investigation, which was only completed on 18 January 1987, when charges were filed against Brukhanov, Fomin, Dyatlov, and Rogozhkin, the head of the shift on the fateful night, as well as against Alexander Kovalenko, the head of the reactor workshop who had approved the program for the tests on the turbines, and Yuri Laushkin, the safety inspector attached to the Chernobyl nuclear power station.

Brukhanov had already been in solitary confinement since his arrest in August. His wife was permitted to bring him a food parcel once a month, and since he had set himself the task of learning English, a British newspaper could be sent in—until a guard saw that his son Oleg had written "Hi, Dad" on one copy, after which this concession was withdrawn. The arrests of Fomin, Rogozhkin, Kovalenko and Laushkin followed. Dyatlov remained in Hospital No. 6 until the beginning of November. A month later, returning from a walk to his flat in Kiev, he was met by two plainclothes officers from the prosecutor's office, who courteously showed a warrant for his arrest and then drove him away in an unmarked car. Akimov and Toptunov, both of whom were dead, were actually sent notification, to their wives' home addresses, that in the circumstances no action was to be taken against them.

The prosecuting team prepared their case with great diligence, working under two constraints. Since the evidence was classified, it had to be cleared by the KGB before they could examine it; second, many of the documents from the fourth unit of the power station remained highly radioactive. Trained as lawyers, they nevertheless had to master scientific and technical details; yet when they wanted to question certain experts, they became aware once again that what they were told had to be approved by the KGB.

To prosecute the case in court, Reykunkov chose his senior assistant, Yuri Shadrin. A plump man with a humorous face that could, at the appropriate moment, turn stern, Shadrin was the son of civil servants and had studied law in Kazan. For the first twenty-five years of his career, he had served as a public prosecutor in Siberia, then he had transferred to Archangelsk and Azerbaijan before reaching his present post in Moscow.

The trial was due to be held in the high court in Kiev, but the venue was abruptly changed to the town of Chernobyl itself, where the auditorium of the House of Culture was transformed into a courtroom. Under Soviet law, the victims of a crime are entitled to attend the trial of those accused of committing it; with 135,000 obliged to leave their homes, as well as the hundreds of victims of irradiation, this would have allowed an uncontrollable number to hear the evidence. By holding the trial in the thirty-kilometer zone, in a building with limited space, the court could give practical reasons for limiting access to the public. The change of order was relayed to the judge, Raimond Brize, from the Politburo by the chief justice, Vladimir Terebilov.

Despite the grandiose title of a judge of the Supreme Court, Brize was in fact a man of lesser stature than the prosecutor, Shadrin; in the Soviet Union, it required some ingenuity to present a tricky case, but none to deliver a verdict that had been decided elsewhere. Thus Shadrin rated a large office with four telephones, Brize a small one with only two. A Latvian from Riga, the son of a railway engineer, Brize was then fifty-five years old. With the bluff, straightforward manner of a worker, he also had to sift through forty-eight volumes of evidence, half of which were moderately contaminated with beta particles, and twelve so highly contaminated that he had to wear protective clothing to read them.

Outside in the streets of Chernobyl, it was hot and dusty. Dosimetrists could be seen checking the levels of radiation in a desultory manner, but none of the groups of liquidators in the town wore masks or respirators. They were advised to wear hats, but no one bothered; despite warnings about beta radiation, there were girls wearing T-shirts and young soldiers with rolled-up sleeves. There were pans of water at the entrace to every building where those who entered were meant to wash the soles of their shoes, and the lawyers were advised to take a shower twice a day (which many did), and change their clothes once a day (which was impractical). Brize and Shadrin had lodgings on the first floor of a vacant building; Brize's two assessors were on the ground floor below. They ate at either the army canteen or that of the Kombinat enterprise that had been formed to clean up the zone. Shadrin was offered a gigantic cucumber that a botanist named Grodzinski from Kiev had grown on contaminated soil. Assured that it was safe, he accepted it; his more cautious colleagues, offered others, refused.

The idea that vodka was an antidote to radiation was widely accepted even by educated people; a week after his arrival, Shadrin was brought a crate of vodka by an old friend, the deputy public prosecutor of the Ukraine, who prescribed a tumbler morning and night. The only one of the judges who refused this preventive medicine died soon after the trial.

7

Shortly before the trial was due to start on 18 March, Nikolai Fomin attempted suicide by slashing his wrists with the lenses of his glasses. Psychiatrists reported to the court that he was in a depressed state and was in no condition to stand trial. Brize decided that the trial could not proceed without him and postponed it until July, when Fomin's mental condition had improved.

The trial began on 7 July 1987, in the auditorium of Chernobyl's House of Culture in the presence of foreign correspondents. Dressed in a dark blue uniform with gold trimmings, showing the two stars of a state prosecutor (second class) on his epaulettes, Yuri Shadrin read out the charges. Brukhanov, Fomin, Dyatlov, Kovalenko and Rogozhkin were all prosecuted under Article 220, Section 2, of the Criminal Code of the Soviet Union, which covered "breach of safety regulations at explosive-prone enterprises or in explosive-prone workshops." Brukhanov and Fomin were further charged under Article 165, Section 2, of abusing their power for personal advantage in knowingly signing a false statement on the radiation levels at the plant and in its vicinity. A further charge of negligence was brought against Laushkin and Rogozhkin under Article 167.

After the reading of the charges on the first day, foreign correspondents were bused out of the zone. Only selected Soviet journalists were admitted to the trial itself. The number of victims who had a legal right of access to the court was limited to ninety-three, of which sixty-seven were from the nuclear power station. However, only twenty-six were in court. They remained under constant surveillance, and any notes they took were later confiscated by the KGB.

The evidence against Brukhanov and Fomin went back to the building of the fourth unit. It was said, in evidence by officials, that the tests on the turbines should have been completed before the start-up, but that Brukhanov had agreed to the units' certification before they were done. Along with Fomin, he was also accused of overlooking breaches in the regulations and of failing to arrange for on-site training for engineers as prescribed by the ministry's instructions. Laushkin was similarly charged with negligence as an inspector. Witnesses described how they had seen workers playing cards on their shifts; there was a general laxity in the plant, "an atmosphere of uncontrollability and irresponsibility under which gross violations of the safety rules were neither prevented nor brought to light." It had been discovered from the logs that in the short period between 17 January and 2 February 1986, the automatic safety systems at the fourth unit had been switched off six times, and Laushkin had done nothing about it.

When it came to the night of 26 April, Metlenko, the turbine engineer from Donenergo, was exonerated because he could not have been expected to know about running the reactor. Dyatlov, who drew up the program, was the principal culprit, but Brukhanov, Fomin and Kovalenko were also blamed for authorizing a program that contained numerous breaches of the rules. A representative of the department of nuclear safety should have been present at the tests; the program should have been approved by the scientific manager (Alexandrov), the chief builder (Dollezhal), the state energy supervisor, and the deputy chief scientific engineer on site (Lyutov), but none had been informed.

The case presented by Shadrin was plausible; undoubtedly there had been carelessness in conducting the tests on the turbines and false statements of the levels of radiation after the accident, but the trial proceeded as if in Soviet society legality was the norm. The actual state of affairs—the arbitrary exercise of power by the party and the necessity for improvization in an economy where planning did not work—was never allowed to spoil the picture of things as they were supposed to be. There could be no admission of failure in the system; therefore the fault had to be with the accused.

In this sense, it was to be the last of the show trials, which had first been staged by Stalin in the 1930s. There was no question of using either torture or blackmail to persuade the five men to plead guilty. All pleaded not guilty to some of the charges, but they were sufficiently creatures of the system to play according to the rules of the game.

The pleas entered by the defense lawyers were mixed. Brukhanov accepted a certain measure of responsibility for the shortcomings in the running of the plant but claimed that the extent of his responsibilities (for the city of Pripyat as well as the power station) had obliged him to delegate such matters to Fomin, his chief engineer. He denied knowing that the fourth unit had been put into operation before certain safety tests had been carried out; however, he admitted signing documents that understated the levels of radiation and failing to initiate contingency plans for the protection of the personnel and the inhabitants of Pripyat.

As presented, this was indeed a heinous crime. In the vain hope that he might somehow conceal the gravity of the accident from his superiors and salvage the award of Hero of Socialist Labor that would come his way when the next two units came on line, Brukhanov had not only left the children of Pripyat to receive a dangerous dose of radioactive iodine but had knowingly condemned members of his own staff to a lingering and painful death—men like Sitnikov, sent by Fomin to inspect the reactor, and humble guards like Klavdia Luzganova, who had remained at her post in the open air.

But everyone knew it was not quite this simple. Whatever his theoretical powers as director of Chernobyl, Brukhanov was a subordinate figure in the hierarchy of the state and the party. Both the party secretary at the plant, Parashin, and the local commander of the KGB arrived at the bunker soon after the director, and by the time Brukhanov signed the incriminating document misstating the levels of radioactivity, regional party leaders had arrived from Kiev. They knew as well as he did the true state of affairs. Brukhanov felt let down at the time of the trial by Vladimir Voloshko, head of the Pripyat town council, and by the two party secretaries, Parashin at the plant and A. S. Gamanyuk in Pripyat, each of whom were merely "rebuked" by the party for the role they had played. However, Brukhanov made no attempt to blame them for what he had done, and Shadrin, who suspected that he had merely followed the advice of the regional secretary in signing the incriminating document, considered that the director had behaved "like a gentleman."

Fomin also confessed to partial guilt. He admitted that there were flaws in the test program that he had approved but insisted that these would not have led to an accident if there had not been subsequent breaches of other regulations. He agreed that both he and Brukhanov should have ensured on-site training for engineers but denied any part in preparing the documents that distorted the levels of radiation.

Dyatlov, too, pleaded guilty to certain technical violations of some regulations but denied that safety had been neglected in drawing up the program. He also denied giving the order to raise the power of the reactor after Toptunov had allowed it to fall; against the evidence of various witnesses, among them Razim Davletbayev, he insisted that he was absent from the control panel at the time. Only the young turbine engineer, Igor Kirschenbaum, who felt that Dyatlov had saved his life by insisting upon his being sent home soon after the accident, said he had not noticed whether Dyatlov had left the control room or not.

Central to Dyatlov's defense, however, was his contention that breaching the regulations was not the cause of the accident. Rather, it had been a fault in the design of the reactor itself, which neither he nor the operators could have known about. The government commission, whose findings were used as irrefutable evidence of the operators' culpability, had no operators serving on it. Moreover, some of its members had the benefit not merely of hindsight but of foresight too—of physically problematic characteristics of the RBMK-1000 reactors that were known to the scientists in Moscow but had never been imparted to the operators who ran them. The documentation, originally drawn up in the 1960s and barely amended since, made no mention of any potential hazard. How could Akimov have known, as the government commission itself had concluded, that by pressing the emergency shutdown button he would actually precipitate an explosion?

Dyatlov's case seemed to be strengthened when, in the course of the trial, it emerged that the reactor had failed to meet the safety requirements on thirty-two different counts. However, in presenting his case, Shadrin ignored the question of the designers' culpability. If there was a case against them, it would be a matter for possible further prosecutions at a later stage. Other RBMK-1000 reactors had functioned for years without serious accidents. He took the line established by Legasov at Vienna: that the accident had been caused by human error. He regarded Dyatlov as a "nuclear hooligan."

Still, the report by the government commission obliged Shadrin to take a somewhat contradictory stand. He depended upon its findings to convict the operators but had to avoid its conclusions about faults in the design, which might have aided Dyatlov's defense. These were referred to simply as "certain peculiarities and shortcomings inherent in the reactor," which played certain "not very well-defined roles." Nor would he allow others to stray into this area; when Nikolai Steinberg, who was called as an expert witness, started to say that the operators could not have known about the positive void coefficient, he was told abruptly by Shadrin that there were no more questions.

Dyatlov, who in the course of the trial had learned much about the RBMK-1000 reactors that he had not known and could not have known before, submitted twenty-four written questions to Judge Brize to be put to the expert witnesses. These were chiefly about the reactor's specifications and whether it met with the requirements laid down by the State Committee for Safety in the Atomic Power Industry. After all, if the reactor had been "prone to explosion," as the prosecution suggested, how could it have been certified as safe? Without providing any explanation, Brize ruled these questions out of order.

Rogozhkin denied the charges against him altogether. As the head of the shift, he had been told about the tests on the turbines but had not been shown a copy of the program. Instead, he and Akimov had referred to an old test program, which had been used when the tests had been tried before. He was not in charge of the tests and had not been informed of any deviations made by those operating the fourth reactor. He agreed that in the wake of the accident he had been told of the high levels of radiation by the head of the plant's civil defense, Vorobyov, but insisted that he had behaved as the regulations prescribed.

Kovalenko also pleaded not guilty. He had given a quick look at the test program, scanning it for about a quarter of an hour, but had seen little in it that affected the reactor. It had not been among his duties to ensure the presence of a representative of the atomic safety department. Laushkin also pleaded not guilty as charged. He insisted that he had performed his duties conscientiously and had been justified in making the assumption that those who were responsible for the test on the turbines would ensure the safe running of the reactor.

Most of the operators who gave evidence at the trial, or who were permitted to attend it, were convinced that their former bosses were being made scapegoats for mistakes that were the fault of the physicists who had supervised the reactor's design. All of them were conscious that they might well have made the same mistakes as Dyatlov, and they had been equally horrified to discover that they had been running such risks for so long. Many of them were also outraged at the way the staff at the

Chernobyl power station were portrayed in the trial as lax and undisciplined. Certainly a witness had come forward to say that he had seen shift workers playing cards or writing letters, but these were not the specialized engineers who actually ran the station; rather, they were superfluous workers foisted off by the state onto the already overmanned plant.

Sasha Yuvchenko was angered by the way technical breaches of the regulations—which in fact were permitted on the authority of a deputy chief engineer and which had no bearing whatsoever on the accident— had been accepted as its cause by those who were technically ignorant. Vladimir Babichev, who gave evidence that Brukhanov had failed to protect his own workers from the high levels of radiation, felt that the real culprit was the scientific deputy chief engineer, Lyutov, who ought to have known about the danger presented by the reactor's design. Nikolai Steinberg was also shocked to realize that all the operators of the RBMK-1000s had been deceived about the risks they ran in the course of their work: the very fact that modifications were under way to reduce the time taken for the control rods to descend into the core from eighteen seconds to four proved that the original design had been at fault.

To the technically ignorant, however, the case against the "nuclear hooligans" was proved. During a break in the trial, Brukhanov's wife, Valentina, went to see Judge Brize in Kiev to get permission to visit her husband in jail. Early for her appointment, she was sitting on a bench in a small public garden when an old man, a war veteran, sat down next to her and started to talk about the trial. People speculated, he said, that they would be sent to prison, but as far as he was concerned, they should be shot.

On 27 July 1987 the court reached its verdict:

> The legal college finds that the defendants Brukhanov, Fomin, Dyatlov, Rogozhkin and Kovalenko are guilty of violations of discipline and regulations guaranteeing safety at plants where there is a potential danger of nuclear explosion, thereby causing human injury and other grievous consequences, in complete contravention of Article 220, Section 2, of the Criminal Code. They find Laushkin guilty of dereliction of duty resulting from a lax attitude about

his obligations, which has caused harm to the government's interests and staff in complete contravention of Article 167 of the Criminal Code.

Brukhanov was also found guilty of abuse of his position for personal advantage under Article 165, and Rogozhkin of criminal negligence under Article 167. The only acquittal was of Fomin on the charge under Article 165 that he had colluded with Brukhanov in falsifying the level of radiation; nevertheless he was sentenced to ten years' imprisonment. So too were Brukhanov and Dyatlov. Rogozhkin received a sentence of five years, Kovalenko of three, and Laushkin of two. Each was ordered to pay costs of 247 rubles and 17 kopeks. Under Article 44 of the law, no appeal was allowed.

PART THREE

RADIOPHOBIA

Ideas that have outlived their day may
hobble about the world for years, but it is
hard for them ever to lead and dominate
life. Such ideas never gain complete
possession of a man, or they gain possession
only of incomplete people.

Alexander Herzen
My Past and Thoughts (1861)

X

1

In May 1988, a little more than two years after the accident at Chernobyl, a conference was convened in Kiev by the Soviet Ministry of Health
and the All-Union Scientific Center of Radiation Medicine of the USSR
Academy of Medical Sciences to discuss the medical aspects of the
disaster. It was attended by over 310 Soviet specialists and 60 from
abroad, including representatives from the International Atomic Energy
Agency and the World Health Organization. Hans Blix, the director
general of the IAEA, addressed the first session, which was opened by
the cardiologist Yevgeni Chazov, who was now minister of health.

Since the first dramatic meeting with experts in Vienna in August of
1986, there had been a second gathering in October 1987, also at the
headquarters of the IAEA. Not only had Gorbachev promised candor
about Chernobyl, but the Soviet delegation at the first conference had
promised to return to answer the four hundred questions that had been
raised by nuclear experts from the West. In the interim, the IAEA's own
International Nuclear Safety Advisory Group (INSAG) had produced a
report based on the information provided by the Soviets. At the October
meeting, although Abagyan had moved toward an admission that the
design of the reactors was not faultless, he had declined repeated requests to go into details. Nor would Ilyn do more than enumerate the
collective dose of radiation received by the population of the Soviet
Union, which, he had insisted, would not cause any sizable change in the
development of cancer.

The meeting in Kiev in May 1988 was altogether different. Ilyn was present, along with Guskova and Baranov, but now the Soviet delegates were on their home ground, and much had happened in the meantime to give substance to *glasnost*. Gone were the Communist rhetoric and Cold War jibes. It was a meeting of scientists and professionals. In greeting the delegates, the chairman of the city soviet spoke of a remarkable milestone, "the millennium of Russian Christianity, the symbol of centuries-old development of our history and culture." Only Anatoli Romanenko, the Ukrainian minister of health, in admitting the "socio-psychological processes that accompany any accident"—that is, the panic—could not resist a dig at the Americans: "We couldn't avoid it in the Chernobyl situation, just as it had not been avoided during the Three Mile Island accident."

Otherwise, the papers presented by the Soviet scientists were strictly professional in tone, giving detailed statistics on the exposure to radiation of the population in the different republics. The gigantic scale of the problem was admitted for the first time: 17.5 million people, including 2.5 million children under seven, had lived in the most seriously affected regions of Russia, Belorussia and the Ukraine; 135,000 had been evacuated from Pripyat and the thirty-kilometer zone. Pregnant women and as many as 350,000 children had been sent to sanatoriums, rest homes and Pioneer holiday camps.

There were some limits to the candor: it was said first that iodine had been given to the whole population of Pripyat within twelve hours of the accident, and later to 5.4 million people, including 1.69 million children. However, Ilyn and his colleagues admitted that "large territories with an increased contamination density by radiocesium existed outside the thirty-kilometer zone. Located in these territories were about six hundred populated areas where people continued to live but lived in conditions of rigid control over locally produced food, including its withdrawal and replacement by clean food brought in from other parts of the country. These territories were called zones of rigid control."

On the whole, the tone of the conference was one of measured self-congratulation. "The Soviet Union," said Health Minister Chazov in his opening speech, "has sufficient potential to solve without assistance the problems emerging from the Chernobyl accident." Romanenko described how 2,000 teams of medical personnel had been mobilized to care for the affected populace, including 7,000 doctors, 2,000 scientists, 1,250 medical students and over 12,000 ancillary workers. By the end of 1986, 696,000 people had been examined, 215,000 of them children. Of these, 37,500 had been sent to hospitals for further investigation, 12,600 of them children.

The number of casualties remained unchanged. Thirty-one had died; a further 209 had suffered from varying degrees of radiation sickness and remained under observation. Of these, 88 percent were able to return to work, but more than 30 percent still suffered from some forms of disablement: heart trouble, gastric disease, respiratory difficulties, sexual dysfunction and disorders of the nervous system. However, all showed signs of gradual recovery, and their disabilities were due to the massive doses they had received.

As to the general population in the affected area, analysis of the radiation situation, and the protective measures taken, "allows us to state with confidence that not a single case of radiation injury has been found in the population exposed to irradiation." Indeed, concluded Chazov, "One must say definitely that we can today be certain that there are no effects of the Chernobyl accident on human health, and that this is due to a great extent to the selfless work of medical specialists, 399 of whom have been given government awards."

2

Among those present at the conference was a scientist named Victor Knijnikov, who, though in no way as prominent as Leonid Ilyn, was foremost among those back-room boys whose research, under the Soviet system, brought honors and renown to the directors of the institutes for which they worked.

Knijnikov was the head of the laboratory at the Institute of Biophysics and was the Soviet Union's foremost authority on the effect of radioactivity in the food chain. Indeed, given the unparalleled facilities for research that had been available to him after the disaster at Mayak in 1957, he was probably more knowledgeable on the subject than any of his colleagues in the West. But if he was less well known than his director, Ilyn, or, for that matter, a man like Legasov, there was little doubt in his own mind that it was because he was a Jew.

Born the son of civil servants in Moscow, Knijnikov had graduated from his secondary school with top marks, but knew that as a Jew it would be futile to apply to a prestigious school of higher education like the Institute of International Relations. Thinking that a doctor could always be useful, he had enrolled in the Moscow medical school and after graduation was obliged to go to Kazakhstan for eight years as part of Khrushchev's drive to cultivate the virgin lands.

There Knijnikov did extensive research on the aftermath of the nuclear explosions around the nuclear testing ground at Semipalatinsk, and later in the contaminated territory around Mayak. In the course of this work, he received a considerable dose of radioactivity but became the country's foremost authority on the absorption of strontium and cesium radionuclides into the body via the food chain, which because of nuclear testing was a matter of acute concern. Knijnikov also did epidemiological research, of great interest to the Americans, on which he reported to an international symposium on strontium held at Chapelcross, Scotland, in 1966. By the time Knijnikov returned to Moscow to work at the Institute of Biophysics, he had twenty-five published articles to his name. When he was appointed head of the laboratory, he joined the Communist party, as was expected. Since he worked in a department of the Ministry of Medium Machine Building, his work was coordinated with the needs of the military-industrial complex and was supervised by the KGB.

Despite his seniority and expertise, it was four days before Knijnikov was told about the accident at Chernobyl; he later suspected that this was to reserve the opportunity to win awards through heroic action for the gentile nomenclatura. Colleagues telephoned him from all over the Soviet Union asking for data that he was unable to supply. It was only at the beginning of May that the Ministry of Health turned to him for advice on contaminated produce, and on 15 May he was summoned to Chernobyl.

Here Knijnikov differed openly with his boss, Ilyn, at a meeting of the government commission. A significant dose of radiation had been received by a village near Bragin in the Gomel region of Belorussia. Children and pregnant women had been evacuated, but the rest of the adult population was overlooked. A month later the question arose as to whether the village should now be resettled. The limit for a single year's dose had been set at ten rems, and it was estimated that the villagers had received nine. Ilyn recommended evacuation, but Knijnikov opposed it. The dose they had already received was from short-lived radioisotopes; the dose they were likely to receive in the future from the long-lived radioisotopes would be in the order of one to two rems, and the trauma of evacuation would do more harm than the extra dose. Knijnikov was overruled. The villagers were evacuated to areas that were later found to be equally contaminated, so they had to be moved again.

The situation facing the experts was vast and complex. Twenty-five thousand square kilometers of land and 2,225 towns and villages were affected—1,845 in Belorussia alone. Their inhabitants, mostly employed in the state-owned and collective farms but also teachers, doctors and local officials, had already been subjected to direct radiation from the cloud of radionuclides that had been spewed out of the reactor. They remained vulnerable to a further dose from the contamination of the soil—through inhaling radioactive dust or eating food that was contaminated by radioactivity on its surface or absorbed from the soil—unless provided with "clean" supplies. Particularly hazardous were mushrooms growing in the forests and the milk from cows that had eaten radioactive grass.

The contamination of the soil was measured in curies per square kilometer, and there was no exact correlation between this and a rad or a rem. Outdoor workers like foresters, herdsmen and farm workers were likely to absorb more radiation from resuspended radioactive dust than factory workers, office workers or schoolchildren. Although the likely dose from radioactive dust was small compared to that from contaminated food, measures were taken to wash down, time and again, dwelling houses and public buildings. Contaminated topsoil was removed to a depth of a few centimeters and buried underground together with the polymer film that had been sprayed from helicopters.

Nothing could be done about the initial, substantial dose of radioactive iodine that had been absorbed by the affected populace other than to administer stable iodine when it became available—in theory as a prophylactic, in practice as a placebo. It had been too late to be effective; examination of the children from Pripyat revealed that some had received doses to the thyroid of 1,500 rads. In the Gomel region, too, the children's thyroids had been affected, but because of the short half life of iodine 131 there was little that could be done other than to keep the affected children under observation.

Once emissions from the reactor had virtually ceased, eleven days after the accident, the principal danger to the populace had been the ingestion of radioactive iodine 131 through milk from cows pastured on contaminated grass. About twenty-five thousand square kilometers of land were contaminated beyond five curies per square kilometer. On 10 May the government commission was given estimates that seventy thousand tons of dried milk would be required to make up for the contaminated supplies. After three months, when the level of iodine had diminished, the danger came from cesium, also ingested by drinking milk. In Kiev, Gomel and other larger towns in the affected area, a ban on the sale of milk could be enforced; in smaller villages where the inhabitants were used to drinking milk from their own cows, such a prohibition was more difficult to enforce. Nor was it easy to prevent them from eating their own vegetables, which were also contaminated, particularly where there were difficulties in importing clean supplies.

If the milk was churned into butter and stored, the iodine 131 content would have decayed by the time it was eaten. It was more difficult to deal with radioactive meat. About eighty-six thousand head of cattle had to be evacuated from the contaminated zone, but because of the shortage of fodder, many had to be killed and buried. The standards imposed were compatible with international norms; the limit for cesium was half the West European norm. In Belorussia Knijnikov himself had to supervise the slaughter of two hundred thousand contaminated chickens. Later the radioactive meat was put into storage to await the decline of the short-lived radioisotopes, and some was eventually distributed to different parts of the Soviet Union, where it was mixed with uncontaminated meat to make a salami-style sausage. Because such sausage was expensive, the theory was that people would only eat small portions, and therefore it would have no perceptible affect on their health. At a time of food shortages, this was considered better than simply burying the meat in the ground. The public was not informed.

By the beginning of June 1986, temporary norms had been established for an acceptable level of radioactive contamination in twenty-four types of food, water and herbs. Teams of radiologists formed by the Ministry of Health carried out hundreds of thousands of checks on milk, meat, fruit and berries, potatoes, vegetables, fish, mushrooms and bread. Where the contamination was over the permitted norms, the food was declared unfit for human consumption. During 1986, up to 10 percent of the fruit and 30 percent of the berries and vegetables were rejected in the affected areas. Milk was often contaminated; only potatoes remained clean.

There were shortcomings in the implementation of such a vast undertaking. Dosimetric instruments were in short supply; teams were inadequately trained; and often there was inadequate coordination between the different services. It was sometimes difficult to impress upon officials and the rural inhabitants the danger of something imperceptible to their senses. Younger men and women, particularly those with children, were quick to grasp the danger and often left the affected zones before they were evacuated. Their parents and grandparents, however, were sometimes reluctant to leave. The hypothetical risks to their health did not impress those who in any case had only ten or fifteen years to live. Some, as mentioned, like Ivan and Irina Avramenko, from the village of Opachichi in the thirty-kilometer zone, hid from the militia and continued to live on the food grown in the private plots behind their homes. Others, having been evacuated, found their own way back along elk paths through the forests. Even families with children eventually returned; often the young ones had been shunned in the villages to which they had been evacuated, treated as radioactive pariahs in their new schools.

Where it could, the militia at first forcibly evacuated even the old, but soon a mixture of compassion and inertia led the local authorities to let them alone. Even in the inhabited zones of strict control, imposing the regulations was a thankless task. It was demoralizing to work in the fields knowing that the food, once grown, would be discarded, and depressing to adhere to guidelines that destroyed the rural way of life. Gathering berries and mushrooms in the forest and pickling tomatoes and cucumbers had been the peasants' pastime in the summer months from time immemorial. How could they be expected to stop doing so now? In winter, cutting down trees, chopping up logs and burning them in their stoves had also been part of the daily routine, but now each stove that burned the contaminated wood became a small version of Chernobyl's fourth reactor. Nor was it easy for the peasants to throw away milk from their own cows and drink instead the powdered product, if they could even find it in the state store. Some herdsmen continued to drink contaminated milk, and on examination were discovered to have accumulated high doses of radiation.

In short, there were innumerable infringements of the regulations and prosecutions of the officials who failed to enforce them. In 1986 and 1987, 23,000 people were fined, 5,500 officials were dismissed and administrative measures were taken against 2,000 more. But the region's inhabitants were demoralized, and some local party bosses used their connections to move out of the zone. So, in time, did some of the resident doctors, themselves ill-nourished and underpaid. In this way many of the villages that had not been evacuated lost not merely the families with children but the few professionals who might have counseled the inhabitants.

Medical teams from Kiev and Minsk appeared from time to time to take blood tests and check these inhabitants' health, and what they discovered was a general level of debility that had nothing to do with radiation. In December of 1986, two of Knijnikov's colleagues reported to the Central Committee and the KGB that the thyroids of children in the Gomel region seemed to be seriously affected. A team of sixty specialists was dispatched from Moscow with orders to investigate the problem and report back in twenty-four hours. It was discovered that there were indeed problems with the children's thyroid, but that these were caused by a lack of stable iodine in the soil, not radioactive contamination.

More pervasive was the evidence of a poor diet for the general populace, and also of chemical contamination far more serious than that caused by radiation. In the summer of 1987, Alexandrov sent a team headed by Sivintsev from the Kurchatov Institute to investigate reports by local scientists that the meat sold in Minsk was radioactive. There they cut a suspect carcass into four and tested it in four different labs, all of which found that the reading came from the natural radioactivity emanating from the potassium in the meat. The level of contamination by nitrates, on the other hand, exceeded the permitted limits by a factor of three hundred.

The brief of the scientific team, however, was not to tackle the question of chemical pollution but to ensure that the food supplied to the people of the controlled zones was free of radioactive contamination. To replace the condemned produce, supplies had to be provided from uncontaminated areas, a considerable logistic undertaking often beyond the capacity of existing channels of distribution. To pay for what they would hitherto have had free from their own plots of land, the inhabitants of the affected areas were given an extra fifteen rubles a month. This became known as "coffin money" and was of little use because there was frequently nothing to buy in the state food stores—a child would be lucky to obtain a single orange each month—and it made people afraid to eat even that local produce that had been pronounced clean.

A considerable effort was mounted to explain the situation by debates, lectures and articles in the local press. To reassure the populace, Knijnikov brought his wife and daughter to a holiday camp in Belorussia in June 1986. The local people brought him small gifts of fruit and fish, ostensibly as a sign of respect for the distinguished professor but in reality to find out whether he would practice what he preached.

Primitive as it was, this testing of Knijnikov's sincerity was not only the best way but the only way for the people in the controlled territories to verify what they were told. True to the tradition of the Ministry of Medium Machine Building and the KGB, all the data on the aftermath of the Chernobyl accident were classified. Not only were readings made by the government's own dosimetrists to remain secret, but the instruments for measuring radiation belonging to the Botanical Institute in Kiev and the Institute of Nuclear Energetics in Minsk were locked away by the KGB, and all the medical records from the controlled zones were taken to the closed atomic city of Obninsk.

The same desire to manage the dissemination of information about the accident was evident in the first days after the accident, when the government medical commission not only released false statistics about the numbers seeking medical treatment but also proposed that imprecise symptoms in the wake of the accident should be ascribed to low blood pressure. Orders were given that all medical information had to be cleared by the government's medical commission before it was printed in the Soviet press. On 27 June 1986, a directive was issued by Shulzhenko, the head of the Third Division of the Ministry of Health, which came under the auspices of the Ministry of Medium Machine Building and the KGB.

The following is to be considered classified:
1. information about the accident at the Chernobyl nuclear power station;
2. information on the results of the medical treatment of the victims;
3. information on the levels of irradiation of the personnel taking part in the liquidation of the effects of the accident at the Chernobyl nuclear power station.

This was not directed at the scientists from the Kurchatov Institute or the Institute of Biophysics, who had long since accepted the discipline of secrecy, but at the many thousands of ancillary workers—among them three hundred army doctors—who had been mobilized in the wake of the accident. Nor was it simply the product of a reflexive secrecy inherited from the Stalin era. There was also the plausible motive that the dangers of radiation were understood only by a handful of scientists like Knijnikov and Ilyn. The chaos in Kiev in early May had taught the government that even local party leaders could behave irrationally as a result of ill-informed rumors.

However, no one knew better than the provincial party bosses that there were times when ideological orthodoxy took precedence over objective truth. The Ukrainian mistrust of Moscow, apparent in the decision to evacuate children from Kiev, resurfaced at the beginning of June, when Ukrainian scientists complained directly to the Central Committee in Moscow that the military dosimetrists—Pikalov's men—were understating the levels of contamination. Their own measurements were considerably higher. Alexandrov was asked to investigate and sent Sivintsev to Kiev. There he discovered that the Ukrainians were using geological radiometers developed to detect uranium, which gave exaggerated readings of gamma radiation. The Ukrainians accepted Alexandrov's explanation, but their suspicions remained. An agreed statement was issued only after considerable negotiation and was a further foretaste of what was to come.

3

Despite the promulgation of the policy of *glasnost*, in 1986 and 1987 the Central Committee still retained the means to impose the secrecy it deemed necessary. It not only controlled the media and the party apparatus but could count on people's residual fear of the ubiquitous KGB, which, though it could no longer arbitrarily arrest Soviet citizens or deport them to Siberia without trial, was nevertheless capable of ruining those who did not toe the party line.

But while the apparatus was still in place, there were some within the Politburo itself who felt that for too long it had been used to conceal the incompetence of state officials. To the apostles of *glasnost* and *perestroika*, the accident at Chernobyl was the fruit of the era of stagnation. Ryzhkov himself was to admit, "We were all heading for Chernobyl." No longer should those who worked under the aegis of the vast Ministry of Medium Machine Building—that state within a state—be allowed to exploit the anxieties of the Cold War to hide their own shortcomings. Within a year of the accident Slavsky was retired, his ministry dissolved and all nuclear reactors placed under the authority of a new Ministry of Atomic Power; but the self-serving inertia in the minds of the populace was more difficult to displace.

The struggle over *glasnost* within the Politburo between Yegor Ligachev and his department of ideology, and Alexander Yakovlev, head of the party's propaganda department, which had started over the reporting of Chernobyl, became more pronounced in the months that followed. This disagreement at the top of the party had serious implications for those farther down, for on the outcome depended the future course of Soviet communism and the choice of political leaders. Innumerable ministers, secretaries, directors and chairmen had much to lose from *glasnost* and *perestroika*; equally, their ambitious rivals had much to gain. Ligachev could count on several powerful groups: the huge military-industrial complex, the powerful bureaucrats in Gosplan, and senior officers in the KGB, but also writers and intellectuals who had a mystical, reactionary and often anti-Semitic vision of Russia. Yakovlev had less powerful and less prominent allies, but he had the all-important support of Gorbachev himself.

The preliminary skirmishes, fought over the reporting of the accident in the newspapers and on television, continued in the arena of the Writers' Union. This body, to which any author had to belong if he wished to see his work in print, was far more than a mere union in the Western sense. With privileges comparable to those given to members of the Academy of Sciences, its members were expected to inspire their readers to follow the party line. If they did not, they were punished—under Stalin with death or imprisonment, under Brezhnev with oblivion. Writers in Russia had the status of prophets, and every political leader was afraid that a Samuel might appear to depose a Saul. Whether in the hands of a Pushkin or a Solzhenitsyn, no ruler in Russia could remain indifferent to the power of the pen.

Yakovlev knew better than most the dangers of alienating the Soviet intelligentsia. In the late 1960s he had served temporarily as acting head of the department he now controlled, and in this capacity had written a ten-thousand-word article attacking the Komsomol journal *Molodaya Gvardiya* for its reactionary, chauvinistic line. Entitled "Against Anti-Historicism," it took various authors to task for idealizing the "stagnating daily life" of the Russian village. This piece had not met with the approval of Ligachev's predecessor as secretary for ideology, Mikhail Suslov, and Yakovlev's appointment was not confirmed; instead he was made Soviet ambassador to Canada, where he remained for the next ten years.

Brought back by Gorbachev in 1983, and later confirmed as secretary for propaganda, Yakovlev used his powers of patronage to install his supporters on the editorial boards of literary journals and publishing houses. Sergei Zalygin became editor-in-chief of the Union's literary journal, *Novy Mir,* and Vitaly Korotich, a Ukrainian, was named editor of the weekly *Ogonyok.* Yakovlev also enlisted well-known writers like Yevgeni Yevtushenko and Andrei Voznesensky to call for more *glasnost* in Soviet literature. This was not welcomed by the leaders of the Writers' Union, frequently talentless time-servers who perceived their privileges as being threatened. In April 1986, the very month of the accident at Chernobyl, the Cinematographers' Union held a congress at which liberals replaced reactionaries in the leadership; among them was Elem Klimov, who had been nominated by Yakovlev himself. The realization that the Writers' Union would be a harder nut to crack led Gorbachev to invite a group of its leaders for a chat in the Kremlin shortly before its own congress opened in June of the same year. There he gave them an impassioned defense of his policy of *perestroika,* and as a graphic image of the kind of inertia that had led to Chernobyl, told them of the sacks of documents on scientific and technical problems that had remained unopened, intended for a special assembly that Brezhnev had at first ordered and then forgotten.

Gorbachev attacked the "layer of apparatchiks" in the ministries and party apparatus, particularly in the monolithic ministry of planning, Gosplan, who served only their own interests. "In our country, nothing is exploited quite so much as official position," the secretary general said. Why was this? Because they were protected from criticism. "We have no opposition. How then are we to control ourselves? Only through criticism and self-criticism, but most importantly through *glasnost.* No society can exist without *glasnost.*"

On this particular evening, Gorbachev's words fell on deaf ears. At the Eighth Congress of the Writers' Union that followed three days later, the old guard either held onto their posts or, as in the case of the first secretary, were replaced by men of the same hue. However, Voznesenksy spoke out uncompromisingly for *glasnost* in literature, and it was clear that in so doing he had the support of at least some of the party's leaders. The reactionaries might be a majority in the Writers' Union, but the liberals seemed in the ascendant in the Politburo.

Soon after the close of the congress, Gubarev wrote his play *Sarcophagus,* and Volodomyr Yavorivsky started his novel *The Star Called Wormwood.* Another Ukrainian, Yuri Shcherbak, while a delegate at the congress, was asked to write an account of the accident at Chernobyl for the magazine *Ionost.* The son of an official in the highways division of the NKVD, the precursor of the KGB, and a party member, Shcherbak was a man of many talents who had won prizes for his films as well as his novels. But he was also a doctor working for the Kiev Scientific Research Institute of Epidemiological and Infectious Diseases, and as such had cared for those evacuated from the thirty-kilometer zone.

Realizing that he would be less constrained by censorship writing for *Ionost* in Moscow than for *Literaturnaya Ukraina* in his home city of Kiev, Shcherbak returned to the zone equipped with a tape recorder, and within a year had completed a book in which different protagonists describe the roles they played in the disaster. Interspersed with these transcripts were expressions of his own feelings, which were similar to those of Yakovlev's department of propaganda. The principal culprits were not the scientists who had designed a faulty reactor, but "Brukhanov, who earlier than others and better than others, understood what had really happened at the station and around it." Shcherbak named the second secretary of the Kiev Oblast Committee, Malomuzh, who had allowed life to proceed normally in Pripyat on the day after the accident, and the directors of the hospital who had known about the high levels of radiation but had done nothing to raise the alarm.

Shcherbak also revealed for the first time that the grandchildren and grandmothers of some senior Communist party officials in Kiev had been dispatched to sanatoriums as early as 1 May while the children of workers were marching in the May Day parade. These wretched "common cowards and scum," relics from the era of stagnation, should be exposed. "It seems to me that the interests of *glasnost*—this very important factor in the restructuring of our society in the spirit of the decision of the Twenty-seventh Congress of the Party—require a fundamental and open review of this problem." To Shcherbak, "one of the most strict lessons of the first month (and of the following ones as well) of the 'Chernobyl era' was given to our mass media, which did not manage to restructure themselves in the spirit of the Twenty-seventh Party Congress. . . . Don't let us mitigate our shortcomings with lies."

To those who knew of the deliberations in the Politburo in the aftermath of the accident, this attack was aimed clearly at Yakovlev's enemy on the Politburo, Yegor Ligachev. But by directing their aim at the commander, the crack troops of *glasnost* inevitably threatened others in their field of fire. If the doctors in Pripyat should have "shouted out loud to the assembly that Saturday morning about the calamity that was drawing near," as Shcherbak suggested, what about Alexandrov, Legasov, Petrosyants, Marin, Mayorets, Israel and Ilyn?

Most of these were sufficiently secure to ignore any implicit criticism of their conduct by a minor novelist in the Ukraine—secure not merely in their posts in the upper echelons of the state but also in their self-esteem and in their view of their own conduct at the time. It was not that they were unaffected by the disaster and the suffering that it had caused, but as Petrosyants succinctly put it at a press conference, "Science must have its victims." In public, all of them subscribed to the idea that the operators had been to blame; in private, those at the Ministry of Energy blamed the Ministry of Medium Machine Building, which had developed the reactor, while those at the Ministry of Medium Machine Building blamed the Ministry of Energy for the poor training of their personnel. The scientists at the Kurchatov Institute blamed Dollezhal's bureau, NIKYET, and those at NIKYET blamed the Kurchatov Institute, but since the investigations into this question promised by the public prosecutor's office had now been shelved and all matters relating to atomic power remained classified, it seemed unlikely that the issue would ever be decided one way or the other.

In the still more private recesses of their own consciences, many of these people undoubtedly suffered from remorse. The two eminent old academicians, Alexandrov and Dollezhal, had seen their life's achievements reduced to ashes in the graphite fire of the fourth reactor at the Chernobyl nuclear power station. Although he had escaped censure himself, Dollezhal had seen his deputy Ivan Yemelyanov punished by the party. Alexandrov, already sorrowing at the time of the accident because of his wife's terminal illness, was stricken by the disaster. Publicly he denied responsibility for what had happened. Even while RBMK reactors were being modified, and both his scientists and Dollezhal's designers were working frantically to develop devices that would ensure that such a mishap could never happen again, he robustly insisted that the RBMKs were entirely safe as long as the operators complied with the regulations.

4

Less sanguine was Alexandrov's first deputy director, Valeri Legasov. Distracted in the aftermath of the accident, first by the Herculean task of bringing the reactor under control, then by the preparation of an ideologically acceptable explanation for the International Atomic Energy Agency conference in Vienna, he had subsequently turned his energetic intellect to a more penetrating analysis of what had gone wrong.

As always, Legasov had great confidence in his qualifications for the task. It was not simply a question of understanding the technical and scientific issues involved; there were others who knew more than he did about nuclear power. Rather it was his unique mix of qualities—scientific, political and psychological—that led to an audacity and an originality denied to others. Throughout his working life, the security provided by his political pedigree had enabled him to take risks that others would have balked at for fear of ending their careers. Of the ten projects he had initiated, five had failed, costing the Soviet state as much as twenty-five million rubles. A further two were still in progress and might possibly fail, but three had proved successful, and one alone had already earned back the twenty-five million rubles with interest.

Legasov was encouraged by the new party policy of *glasnost* and *perestroika* to think more radically than might have been wise before, and even to offer a certain measure of self-criticism. He felt a small twinge of bad conscience about his role in blaming the operators for the disaster; "I told the truth in Vienna," he told a friend, "but not the whole truth." Yet he knew as well as anyone how difficult it had been to be allowed to say anything at all. More troubling was the part he had played in the years before the accident to counter any criticism of nuclear power. As Alexandrov's first deputy director, it had come under his theoretical jurisdiction and had been included in the inquiry he had conducted into the whole question of industrial safety. It seemed incredible now that at that time nuclear reactors had seemed to pose less of a threat to human health than the fossil-fuel power stations that spewed large quantities of carcinogenic chemicals into the atmosphere.

Legasov remembered feeling a certain unease when he had considered the RBMK reactors: as a chemist, he was disturbed that their design incorporated large quantities of graphite, zirconium and water. He was also uneasy about the safety systems and knew that modifications had been proposed that, because of the cost of their implementation, the designers had neither accepted nor rejected but left pending. Why had he been so complacent? Why had he been so emphatic in supporting Alexandrov's contention that there was nothing to fear from nuclear power? Because it had never occurred to him that men as eminent and experienced as Alexandrov and Dollezhal could be mistaken. The discussions he had had with Alexandrov were conceptual, not technical. Criticisms had been made of the RBMK reactors on the grounds of their cost and efficiency, not of their safety.

Why had they been so confident that everything would go according to plan? Undoubtedly the excellence of the scientists working at the Kurchatov Institute had lulled them into a false sense that the same quality was to be found among the operators and engineers. Looking back, he now remembered the instances of shoddy workmanship that had come to light—for example, the flawed seams in the piping where welding, although certified, had never taken place. But what had shocked him more than almost anything else at Chernobyl had been the tape of the operators' conversation prior to the accident, which he now kept locked in his safe. Seeing that a section of the program for testing the turbines had been crossed out, one had telephoned a colleague to ask what he should do. The second operator appeared to hesitate, then said, "Follow the deleted instructions." All the while, the safety supervisors were on the site, unaware that the tests were going on.

The personnel had been punished; had they survived the accident, Akimov and Toptunov would have joined Dyatlov and Rogozhkin in jail. Yet Legasov knew that he had done them an injustice at Vienna by suggesting that they alone were to blame. There were no specific culprits for what had happened at Chernobyl. There were failings: the designers had dragged their feet about modifying the safety systems; the operators had developed casual procedures, neglecting the regulations. But when he looked at the precise chain of events and at the particular motive for each action, it was impossible to define the sin. If the operators made mistakes, it was because they considered it a matter of honor to complete the tests on the turbines. The accident, then, was not the consequence of dereliction of duty, but rather of going beyond the call of duty to finish the job on time. It was not so much an isolated case of negligence as an attitude of mind, the apotheosis of all the inefficiency and incompetence that had bedeviled the Soviet economy in recent decades. Truly, as Ryzhkov had said, the nuclear power industry had been moving toward this terrible event with a certain degree of inevitability.

Before he went to Chernobyl, Legasov had suffered from a recurrent kidney disease. His heroic work at the reactor—sleeping only a few hours a night, subject to constant stress and significant doses of radiation—had not improved his condition, but with demonic energy he shrugged off this ill health to work on his inquiry into the causes of the disaster. What caused the slovenly habits that had become endemic to

Soviet life? He thought back to the heroic age of Soviet nuclear power when Kurchatov, Alexandrov and Dollezhal had astonished the world with their achievements. What had changed? Only the size of the enterprise. To have a single group of men designing and building a reactor was fine when it was small, but once each had his own institute employing tens of thousands of scientists and engineers, accountability was lost in a maze of departmental and interdepartmental committees. Collective responsibility meant, effectively, that there was none.

What was to be done? Huge monopolistic structures like the Ministry of Medium Machine Building must be split up into smaller entities competing in a form of internal market. In the construction of nuclear reactors, there must be a clear distinction between supplier and client. So too, huge institutes like the Kurchatov, employing ten thousand scientists, all for life, and many working on projects of dubious validity, should be broken up into smaller units, each with its own scientific leader, who would employ scientists for particular research projects and for only a limited period of time. Moreover, the state should become more discriminating in its allocation of funds. No longer should a block grant be paid to institutes like the Kurchatov simply because they were there. Nor should funds be allocated by old men who held their posts by virtue of their appointment by the party. Youth must take the helm. Each team should have to justify its expenditures, and the millions being spent on prestigious but essentially fanciful schemes like Velikhov's research into nuclear fusion might be reallocated to less dramatic but more urgent projects, such as decontaminating the territory around Chernobyl, ensuring the safety of nuclear reactors and protecting the environment.

Confident that his analysis was correct and impatient to implement the reforms he proposed, Legasov presented his report to the Academy of Sciences. He had good reason to think that it would be well received. As a convinced Communist, as well as a scientist of great distinction, he had devoted his life to the service of the party. He had risked his life at Chernobyl, his reputation at Vienna and had now risen to the challenge of *perestroika*.

Legasov's report was neither accepted nor rejected: it was ignored. A coalition of forces combined to put his proposals for reform in limbo.

His fellow scientists at the Kurchatov, who saw their own projects placed in jeopardy by the proposed reallocation of resources, were outraged. Legasov was not a physicist but a chemist. Like many others who had led comfortable and privileged lives under the aegis of the Ministry of Medium Machine Building, these men were unwilling to lose their sinecures without a struggle. Alexandrov, despite his affection and respect for Legasov, was implicitly blamed by the report for the shortcomings of the status quo. L. D. Riabev, who had replaced Slavsky as the minister responsible for atomic power, while sympathetic to Legasov's point of view, thought his proposals went too far and too fast.

Nor were the advocates of *perestroika* willing to let Legasov, the favored protégé of the Central Committee, change horses in midstream. His eminence had aroused envy. Rather than being regarded as the hero of Chernobyl, he was seen by many of his colleagues as a typical product of the era of stagnation, a man who owed his position to the kind of political nepotism that had led to the disaster in the first place. There were also the Jewish scientists, who felt that Legasov had discriminated against them when allocating research funds or authorizing trips to the West.

All these various forces opposing Legasov were given a chance to humble him when, in the spring of 1987, as a measure to foster *perestroika*, the Central Committee ordered that there be elections to a supervisory Works Council in all Soviet enterprises and institutions. Too late, Legasov realized the dangers he faced if he were to run. Despite the influence he might possess in the Politburo or the secretariat of the Central Committee, he could hardly be appointed director of the Kurchatov Institute in succession to Alexandrov if it were shown that he did not enjoy the confidence of the 229 eminent scientists who were qualified to vote.

Pleading ill health, Legasov told Alexandrov that he would prefer not to be considered as a candidate for the council, but his mentor insisted that he run; it was impossible for the first deputy director not to do so. Other colleagues agreed; even his enemies encouraged Legasov, nurturing the pride that comes before the fall. When the vote was counted, 100 were for Legasov and 129 against.

Despite his premonitions, Legasov was dumbfounded by this evidence of his unpopularity. It was the first setback he had ever encountered, but not the last. Some weeks later, on 10 June, after a meeting of the party cell at the Kurchatov, Alexandrov told his colleagues that they should congratulate Legasov. He had seen the final list of those to be decorated for their heroic deeds at Chernobyl. The highest award, Hero of the Soviet Union, was reserved for the three commanders, Teliatnikov, Antoshkin and Pikalov, but just beneath them, and deservedly so, came the name of Legasov. "Tomorrow our first deputy director will be proclaimed a hero of socialist labor," Alexandrov said.

The congratulations were premature. When the list was published, Legasov's name was not on it. At the last moment, it was said, Gorbachev had thought it inappropriate to honor any of the scientists from the Kurchatov for their role in a disaster for which they were partly to blame.

The next day, Legasov telephoned his secretary from his home, and at the end of their conversation asked her to keep an eye on his two children. This strange request made her fearful and she raised the alarm. Legasov was found semiconscious, an empty vial of sleeping tablets by his side. An ambulance took him to the hospital, the drug was pumped out of his stomach, and he survived.

5

Legasov returned to his duties at the Kurchatov a broken man. He was isolated by his high office; even the friends of his youth could not help him. Nikolai Protzenko, who had once taught Legasov and had been with him in Tomsk in the 1960s, continued to admire him as a scientist and felt affection for him as a friend. But he had become estranged from Legasov, the "scientific leader," who had surrounded himself with secretaries and hobnobbed with members of the Central Committee. The old Russian saying, "You can come in without knocking at the door" had long since ceased to apply.

As Legasov's spirits sank, his health declined. Arriving one day at the entrance to the administrative block, Yuri Sivintsev found him shuffling up the stairs as if he were as old as Alexandrov. All his zest seemed to have left him, even on those coveted trips abroad. Attending a conference in London that summer, Legasov ran into Gubarev, who had come to see a performance of his play *Sarcophagus* at the National Theatre. To try to raise his spirits, Gubarev suggested going to the musical *Cats* or finding a couple of girls, but Legasov preferred to return to his hotel.

In October, Legasov went into the hospital to be treated for his kidney disease. Alerted by Margarita Legasov to her Faust's unhappy condition, Leonid Ilyn visited him at his bedside. "Leonid Andreevich," Legasov asked him, "how is it that after Chernobyl I got dozens of invitations from different countries to go and work there with my family—they offered me a house, a car and every other Western luxury—but here in Russia I am frustrated at every turn. No one supports me within the institute, the ministry or the Academy of Sciences. I am ostracized." Ilyn could not answer. In his own institute, he may have had enemies, but he also had a group of friends and colleagues who had been with him from the beginning and would stand by him to the end.

Tormented by his thoughts and unable to sleep, Legasov took another overdose of sleeping pills, but again was saved by doctors. News of what had happened was leaked to the press, which, if it was not yet free in the Western sense, was capable of Fleet Street innuendo. The magazine *Moscow News*, an advanced proponent of *glasnost*, published an unsigned interview with Legasov and beneath it an article entitled "Give Me Poison."

In an attempt to revive Legasov's interest in life, Gubarev asked him while he was still in hospital to develop his ideas on industrial safety in an article for *Pravda*. Legasov went to work with his habitual zeal and finished the article in a week. Every day after it was published he telephoned Gubarev for a report on the reaction to what he had written. There was none.

Stung by official indifference to his warnings, Legasov became more strident in what he said. From his hospital bed he gave an interview to a Belorussian writer and filmmaker, Ales Adamovich, for the liberal journal *Novy Mir*, warning him that "another Chernobyl could happen at any station of that type, in any sequence."

Astonished that Alexandrov's deputy would make such an allegation, Adamovich asked if he could tape the interview, and Legasov agreed. "It's not so easy," he went on, "to completely eliminate the major components of the Chernobyl catastrophe. They include the fact that no principles for creating completely reliable emergency systems for such stations have yet been developed. They include the impossibility of building containment structures over them, even now. Therefore, a repetition of what happened is not ruled out. . . ."

"Do other scientists understand this?"

"Many do."

"Why don't they tell the government?"

"Apparently the old notions are still at work. No one will listen if they are told since they follow the official line. The inertia, the orientation toward atomic stations no matter what, is too great. We began their wholesale construction ten years late, and so we set off at a gallop, not fully equipped, economizing on containment structures and other things, just so that there would be a little more money. It's hard to stop. But the main thing is that behind all this are departmental interests, the interests of individuals, officials and groups, including the interests of various scientific sectors."

From his office in the Kurchatov Institute Legasov also spoke to the Ukrainian writer Yuri Shcherbak, who was gathering material for his book on the disaster. Now Legasov's ruminations took him deeper into the tormenting conundrum of Chernobyl. Certainly, Legasov thought, as the Politburo's commission of inquiry had established, there had been "a certain incompleteness, even slovenliness in our work. At all stages, from the creation to the running. . . . And all the time I thought: why does this happen? And, do you know, I come to a paradoxical conclusion . . . that this happens because we have got too carried away with technology. We have become too pragmatic with naked technology. This embraces many questions, not only of safety. Let us think for a moment: when we were far poorer, and the international situation was far more complex, why in a historically short period, from the thirties to the fifties, did we manage to astonish the whole world with the rate of creation of new types of technology and be admired for its quality?"

What Legasov had come to realize, he told Shcherbak, was that "the technology of which our people are proud, which ended with Gagarin's flight, was created by people who stood on the shoulders of Tolstoy and Dostoyevsky. . . . The creators of the technology of that time were educated in the spirit of the greatest humanitarian ideas. In the spirit of beautiful literature. In the spirit of great art. In the spirit of a beautiful and correct moral sense. And in the spirit of a clear political idea of the structure of the new society, the idea that this society was the most advanced in the world. This high moral sense was there in everything: in the attitude of one person to another, in the attitude to man, to technology, to one's duties. All this was there in the education of these people. And technology for them was simply a means of expression of moral qualities, placed in them . . . which Pushkin, Tolstoy and Chekhov taught them to have toward everything in the world."

"For a prolonged period," Legasov went on, "we have been ignoring the role of the moral principle, the role of our history and our culture." What was the result? A generation of young men and women who, like the operators at Chernobyl, were technically proficient but morally indifferent. "We will not cope with anything," he told Shcherbak, "if we do not renew our moral attitude to the work that is being done, whatever sort of work it is—medical or chemical, biological or to do with reactors."

"But how can this moral attitude be renewed?"

A sigh, then after a long pause, "Well . . . I can't be a prophet."

6

Well aware by now that there was little chance that he would succeed Alexandrov, Legasov drew up a plan to set up his own institute of industrial safety and submitted it to the Academy of Sciences. There was a good chance that it would be approved. Even if they could not be acknowledged by a decoration, his heroic achievements at Chernobyl must surely count for something, and his patron, Yegor Ligachev remained second only to Gorbachev in the hierarchy of the party.

A decision was to be made at a meeting of the Academy at the end of April. In March, however, Ligachev staged a rear-guard action against the forces of *glasnost* and *perestroika*. Taking advantage of Gorbachev's absence abroad, he authorized the publication in the Russian nationalist paper, *Sovetskaya Rossiya*, of an apparently unsolicited letter from a chemistry teacher in Leningrad, one Nina Andreyeva. Entitled "I Cannot Give Up My Principles," it attacked the program of reforms as being inspired by Western values and implemented principally by Jews. Further, the attacks on Stalin's reputation, which Yakovlev had authorized, were feints against communism itself, and the proponents of reform were, like the Jews Trotsky and Martov, covert enemies of Lenin and the October Revolution.

For a month thereafter, it remained uncertain whether this outspoken attack had rallied the conservative opposition on the Central Committee. On 15 April, however, *Pravda* published a detailed rebuttal of the charges made by Nina Andreyeva entitled "The Principles of Perestroika." Said to have been written by Yakovlev himself, its anonymity enhanced its authority. The party had spoken, and Ligachev was discredited. In falling back upon chauvinism and anti-Semitism, the secretary for ideology on the Central Committee, the Grand Inquisitor of world communism, had exposed both the bankruptcy of his beliefs and the limits of his power.

Ten days later, on 25 April, a meeting of the Academy of Sciences considered Legasov's proposal to set up an autonomous institute for industrial safety. Alexandrov's support was lukewarm, and it was turned down. Legasov learned of the result on the evening of 26 April, two years to the day after the accident at Chernobyl. The next day, when Legasov's son returned from work he found his father's dead body hanging from the balustrade in the stairwell of their home, a rope around his neck.

At the conference in Kiev a fortnight later, Academician Ilyn was congratulated on the award he had received to mark his sixtieth birthday a few days before the death of Legasov; he was now a Hero of Socialist Labor. Legasov was not forgotten. In his address to the participants, Hans Blix, the director general of the International Atomic Energy Agency, spoke of "the excellent example set by the Soviet Union when, after the Chernobyl accident, it produced an extensive report on the causes and course of the accident and submitted it to international discussion with some five hundred nuclear experts in Vienna in August 1986. We regret today the untimely death of Valeri Legasov, who stood with several Soviet colleagues at the center of this discussion, which enabled the nuclear world to learn from the tragic experiences of Chernobyl."

At the Kurchatov Institute some weeks later, Yuri Sivintsev asked his director, Academician Anatoli Alexandrov, who should now sign the letters sent out by the Council on Ecology, which Legasov had set up before he died. Tears came into the old man's eyes, and he embraced Sivintsev and sobbed. "Why did he abandon me?" he asked. "Oh, why did he abandon me?"

XI

1

Less than a month after Legasov's death, Gubarev published the scientist's memoirs in *Pravda:* a brief account of what he had done at Chernobyl, which he had dictated into a tape recorder, followed by an analysis of what had gone wrong. "Valeri Alekseyevich's departure from this life is hard to explain or understand," wrote Gubarev in the introduction, "why at the height of his powers he should kill himself. This tragedy should be a lesson to all of us and a reproach to those for whom tranquility and well-being come first." He quoted the panegyric of Legasov's fellow academician, Yuri Tretyakov: "Legasov is Don Quixote and Joan of Arc at the same time. An inconvenient and difficult person for those around him, but without him you have a sense of emptiness and loss of something close to the meaning of life."

By publishing this memoir, and Legasov's conclusion that "Chernobyl was the apotheosis, the summit, of all the incorrect running of the economy that had been going on in our country for many decades," *Pravda* at least ensured that the man had not died in vain; the voice from the grave unequivocally endorsed *glasnost* and *perestroika.*

However, Legasov's posthumous admission of such dire inadequacy in the nuclear industry not only supported the case for political reform but also justified the increasing alarm of the population about the safety of other reactors. As early as August 1987, a letter was published in *Literaturnaya Ukraina* from seven Ukrainian writers protesting the construction of another nuclear power station on the banks of the Dnieper near the city of Chigirin. This project had a background of particular incompetence, but the significance of their letter was not simply that it was published, or that it revealed yet another fiasco from the era of stagnation, but that the writers included Vasili Zakharchenko, a former dissident who at one time had been imprisoned in a labor camp in Perm, as well as Fedir Morhun, the first secretary of the Poltava District Party Committee. Former ideological adversaries had sunk their differences to ask, "Can it be that the Chernobyl tragedy has taught us nothing?"

Exacerbating these fears of a second disaster were numerous unofficial reports of sickness among those who had been exposed to radiation at Chernobyl. Rumors spread that many of the six hundred thousand men and women who had worked as liquidators in the thirty-kilometer zone had either died or fallen ill. In 1987, an association, the Chernobyl Union, was formed to provide aid and information for all those who had suffered from the accident. At once, complaints were received from its members about chronic ill health. For example, Eduard Korotkov, one of the pilots of the experimental Ka-27 helicopter, suffered first from appendicitis, then peritonitis and finally gangrene of the gall bladder. Recovering from the last operation, he had a heart attack. Suspecting that exposure to radiation had damaged his immune system, the surgeons sent him to be examined at Moscow's Hospital No. 6. There Guskova insisted that his maladies had nothing to do with radiation.

Vladimir Lukin, one of Antoshkin's helicopter pilots, who had flown in from Torzhok, received an official dose of eighteen rems before leaving the zone. Upon his return, he suffered from repeated headaches and became so easily tired that he was obliged to retire from the air force. Other pilots, veterans of Afghanistan, suffered in a similar way but concealed their symptoms to keep on flying and qualify for a full pension.

Among the six hundred metro construction workers, forty became invalids within five years, suffering ailments as different as kidney failure and heart disease, which they ascribed to radiation. Four died, including the metro construction workers' chief engineer at the time of Chernobyl, Victor Koreshkov. Ylena Holod, a physician treating the liquidators still working in the zone in the clinic at Chernobyl, thought she noticed a marked increase in stomach disease, cardiovascular disorders, arteriosclerosis and premature aging. Unnatural lethargy and sweaty palms were clear signs to her of a decline in the number of lymphocytes in the blood.

This diagnosis made by the doctors in the field was dismissed by the specialists in Moscow and Kiev. The twenty-five rems that had been received by the liquidators could not conceivably lead to any form of radiation disease. What people suffered from, said Academician Ilyn, was "radiophobia," a psychological condition whereby every illness, from eczema to the common cold, was ascribed to radiation. If there was a deterioration in people's health in the contaminated zones, it was possibly because people were afraid to drink the milk or eat the vegetables that had been imported by the authorities, and had thereby deprived their bodies of essential nutrients.

Even as Ilyn reassured the populace, his scientists studied samples of blood for deviations from the norm, for though the liquidators had been limited to a dose of twenty-five rems, dosimetric controls were often inadequate and sometimes inexact. It was easy enough to measure background radiation, but there were many variables—small "hot spots," for example, of much higher levels of radiation, or gusts of wind that could blow radioactive dust that entered the lungs of the workers. They did not find any evidence of damage to the immune system, but they did discover that radiation appeared to have affected the reproductive organs of some of the young soldiers, leading to "lazy" sperm. It was this that had led to the decision to replace the young conscripts with the older "partisans."

The research of Ilyn's scientists remained classified: there was no opportunity for the physicians working for the republican governments to verify the findings of the Institute of Biophysics in Moscow, or even the new All-Union Scientific Center of Radiation Medicine, which had been set up in Kiev. All the medical records of the liquidators were kept by the Third Division of the Ministry of Health in the closed city of Obninsk. Lubov Kovalevskaya, working as a journalist in the zone, tried to establish a register of her own but discovered that orders had been given that service in the zone was not to be entered in a soldier's record.

It was the same for those who still lived in contaminated territory. Based on the data supplied by Israel's hydrometeorology committee and Knijnikov's laboratory at the Institute of Biophysics, the government commission had divided the contaminated areas into four zones:

1. a zone of "alienation," from which the entire population had been evacuated;
2. a zone of temporary evacuation, in which, against official advice, one thousand people had returned to live and refused to move;
3. a zone of constant control;
4. a zone of periodic control, in which eighty-four thousand people lived in 176 villages. Of these, forty-seven thousand lived in areas where contamination of the soil by cesium 137 was over fifteen curies per square kilometer, which meant that food had to be brought in from outside.

This strategy was not made public until 1989. In the style of Russian governments established well before the Revolution of 1917, orders issued from Moscow were expected to be obeyed without question or complaint. But this did not prevent the doctors who worked in the controlled zones from reporting an unusual amount of ill health. There were not only cases of thyroid dysfunction among the children—this had been expected—but also reports of cataracts, an unusual increase in bronchial and gastric infections and vague symptoms such as lassitude, headaches and depression.

Again the government ascribed these to "radiophobia," not merely a fear of radiation but also a misunderstanding of what precautions should be taken in the contaminated zones. The father of a child with rickets told Professor Knijnikov that he had made sure that his son had drunk no milk for the past three years. Moreover, the very thoroughness of the medical checkups in the contaminated areas uncovered a high level of endemic ill health. It was difficult to make comparisons with people's conditions before the accident because no satisfactory statistics existed as a base against which to measure the new figures. Moreover, those that did exist were often false; epidemics of influenza, for example, had been concealed by ascribing the symptoms to a variety of different diseases.

There were other factors. Because younger couples with children had left the area, the average age of the population had increased, with a consequent decrease in the ratio of healthy to unhealthy people. Where children remained, the danger of walking in the contaminated forests and the danger from eating mushrooms and berries had often resulted in parents keeping their children indoors, thereby depriving them of fresh air and their traditional sources of vitamins. Where they had been resettled, the difficulties they faced as evacuees, with no homes and no jobs, often billeted in frightened and unsympathetic communities, caused acute stress. The inhabitants of Pripyat, where life had been so agreeable and easy, suffered particularly from being so abruptly and brutally ejected from their homes. All this was undoubtedly a source of suffering, but could not be ascribed to the biological effects of radiation.

2

In August 1987, three months after the conference at Kiev at which the Soviet medical authorities had apparently come clean about the consequences of the Chernobyl disaster, the reassuring scientific consensus was destroyed by an interview published in the Soviet magazine *Nature* with the distinguished Ukrainian botanist Professor Dmitri Grodzinski.

A biophysicist, radiobiologist and member of the Botanical Institute of the Ukrainian Academy of Sciences, Grodzinski had been ordered by the Ministry of Agriculture immediately after the accident to form a team to study the effects of large doses of radiation on the resistance of plants to disease. There was a fear that if this resistance decreased, it might lead to a widespread crop failure like the nineteenth-century Irish potato famine. The team began experiments with wheat and other plants within both the ten- and the thirty-kilometer zones, and in a greenhouse in the abandoned town of Pripyat, which produced some interesting results. It was found that while pine trees, like human beings, could be killed by a dose of six hundred rads, birch trees could withstand ten thousand rads and tomatoes thirty thousand. Indeed, some plants flourished after a dose of radiation; it had been one of Grodzinski's outsized cucumbers that was offered to the public prosecutor, Yuri Shadrin at Chernobyl.

Despite the extension of his botanical research into the classified areas of radiobiology and biophysics, Grodzinski's institute in Kiev was neither controlled nor funded by the Ministry of Medium Machine Building. He had none of the instinctive military discipline of an Ilyn; preferring the novels of Walter Scott to the collected works of Marx and Lenin, he was in appearance and manner the classic absentminded professor, with rimless glasses and thick tufts of hair surrounding his bald dome. The blackboard on the wall of his office was covered with different formulas, graphs and theorems, and piles of unsorted papers lay on his desk.

At the time of the accident at Chernobyl, Grodzinski's laboratory in Kiev had registered a sharp increase in radioactivity. Instead of being authorized to warn the city's people and suggest precautions, he had been told that his knowledge was classified, and his instruments had been impounded. In vain he had recommended iodine prophylaxis for the populace. He was told that this would only lead to panic, even though fear was already leading people to drink iodine intended for disinfecting wounds. Later he had suggested adding calcium to bread to make up the deficiency for those in the controlled zones who were afraid to drink milk. This suggestion, too, was never adopted, nor were his anxieties made public. On the contrary, in an optimistic interview with TASS in June 1986, Grodzinski had described the different ways in which the soil around Chernobyl could be either purified or used to grow crops unaffected by radiation:

> Chernobyl soils are quite suitable for seeding perennial cereal grasses, such as English blue grass, orchard grass, and others. The fact is that radionuclides have virtually no effect on them. Following the chemical amelioration and a reliable consolidation of radionuclides deep into the ground, the land will return to normal, full-blooded life. Fields will no longer be quiet, and machines and people will appear again. Life will again settle in its habitual tracks. But it will be necessary to go to some trouble to restore this land to life.

Two years later, in his article in *Nature,* Grodzinki changed his tune:

> It needs to be said openly that specialists today still know relatively little about radiation. That is why a large body of radiobiologists tend towards the view that even the smallest dose is harmful. . . . There is one important fact to grasp in connection with the collective dose of radiation. The collective dose is the sum total of the radiation of all the population. It is a delayed-action bomb. It seems

to be the case that the large part of the injury done to the population is in the form of hidden recessive mutation. Mutational defects of this kind can pass into the homozygotic makeup, and hence manifest themselves as different genetic defects in the families of succeeding generations.

The likelihood of these defects appearing was the so-called "coefficient of risk," but because of the uniqueness of the Chernobyl disaster, any kind of accurate prediction as to its consequences was impossible:

If we are talking about us getting used to it [the collective dose of radiation] in some way, then we are talking about a process which would take a very long time—the whole of an evolutionary period—and the end result of the radiation would still be the death of the population.

The cause of this genocide, Grodzinski suggested, was not just the accident itself, but the failure of the authorities to warn the population about what had happened. Describing how his instruments had been impounded at the time, he blamed the officials "who were ignorant but had the power":

Secrecy and ecology do not mix. This secrecy springs from non-professionalism. When the accident happened, I met many of these so-called "professionals" who were expounding verbiage about the accident and were giving recommendations, although they didn't possess the slightest idea of the collective dose I have told you about or anything about the stochastic [probabilistic] effects of radiation.

Grodzinski contrasted this silence with the action of the Finns, who only a few days after the accident "had already published . . . in black and white, how and what people could do within the polluted territory . . . where children can walk, how much and in which regions cattle should graze, what to eat, what to drink."

With no similar guidance in the Soviet Union, people had become afraid to drink milk, thereby depriving themselves of necessary calcium.

As a result, two years later, "Kiev is gripped at the moment by an absurd speculation as to what was and what will be. It is gripped by fear. On television they are calling for society to remain calm. On these programs, there is one leitmotiv: that there is nothing to be done to make the situation better; that the situation is getting better and will continue to improve all the time; and that the ecological consequences are zero or practically zero."

What was to be done? "First," suggested Grodzinski, "The population ought to be provided with individual dosimeters, like the Japanese who go to market armed with these devices so that they can check the radioactivity level of their cabbages and fish."

Further, people should also be advised on and provided with a vitamin-rich diet to bolster their immune systems. "People in general stopped eating vegetables because they were afraid of radiation. This is simply incorrect. In an organism deprived of vitamins, radiation acts all the more strongly."

Grodzinski did not recommend alcohol, which, though it increased resistance to radioactivity by 1.13 times, also destroyed the vitamin molecules that acted as "radioprotectors."

The most important lesson to be learned, however, was the need for *glasnost*. "Newer and newer problems keep revealing themselves all the time, problems which link up to such different fields of knowledge as psychology, medicine, radiology, mathematics, and construction. We shouldn't expect that the number of these problems will suddenly and dramatically lessen. Quite the opposite: they will increase. I will underline this point once again; we must not, on any account, remain silent. We need *glasnost*. Each secret does more harm than good because it conceals the problem and prevents the academic community from arriving at a solution."

To illustrate the article, Grodzinski produced photographs of strange leaves and pine needles—instances of gigantism found within a few kilometers of the zone. In some cases the leaves of oak trees, normally twelve centimeters in length, had grown to seventy centimeters; and the leaves of linden trees to forty centimeters in diameter. In Grodzinski's opinion, this was caused either by radiomorphosis or mutation; in subsequent generations of wheat grown in the ten-kilometer zone, interesting new species had been developed—one of the few gains from the accident at Chernobyl.

The readers of *Nature,* however, reacted less calmly to what Grodzinski had revealed. The issue in which his article appeared quickly sold out, and copies started to change hands at fifteen times the published price. It was reprinted in other papers, and broadcast from Munich over the Voice of America. The photographs of freak leaves, taken with the view Grodzinski had expressed that even the smallest dose of radiation was harmful, and talk of the "death of the population," caused widespread alarm. Thousands abandoned their homes and fled from the controlled zones without waiting for official sanction.

Worse was to come. To the horror of the inhabitants, a sow in Narodici gave birth to a litter of piglets without eyes. News of this spread, and further freaks were discovered in the same region: a foal with eight legs, a chicken with a dragon's head, a piglet with an eye half the size of its head, a calf with a lip like an elephant's trunk, a goat with its hind legs three times longer than its front ones. Local veterinarians in Narodici reported that these genetic abnormalities had dramatically increased since the accident at Chernobyl. A small museum was founded with photographs of the freaks, which foreign newspapers were only too happy to copy. Vivid photographs of them appeared in newspapers and magazines abroad, and pictures of deformed children accompanied horrifying stories from the Ukraine. In London's *Sunday Times* it was reported that "hospitals in the Ukraine, Byelorussia and adjacent provinces of Great Russia are filled with victims. Whole wards are lined with gaunt, dying and deformed children."

In vain the Ministry of Health denied any increase in abnormal babies. Even Grodzinski was shocked when he saw what he took to be photographs of thalidomide-affected children portrayed in foreign magazines as the victims of Chernobyl, while Professor Vorobyov in Moscow denounced the stories as lies. However, Grodzinski was more ambivalent about the defects in animals. Although he had seen numerous mutations in plants within seven kilometers of the reactor, he thought it would be impossible to prove that the deformities in piglets were caused by radioactivity, but he did think it possible that if a pregnant sow lay on hot particles, they might penetrate the skin and be carried by the blood to the placenta. All the cells around the hot particle would die, and this would cause deformities. However, he was willing to admit that no one else supported his theory, and that in not doing so "they might be perfectly right."

Even mutations of this kind, however, were not to be confused with genetic abnormalities. In Grodzinski's opinion, a genetically malformed fetus would be recognized by the body and aborted. However, neither the reassurance of the specialists nor the insistence of experts from Agroprom that these animal deformities came from inbreeding or pesticides in the feed, and were also found in uncontaminated areas, calmed the inhabitants of the blighted areas. Many were convinced that the defects were caused by radiation, and there was a sudden spate of human abortions.

Victor Knijnikov, Ilyn's head of lab, who had studied the effects of radiation on the food chain for the past twenty-five years, was at first amused at a botanist's excursion into his field, but when it became apparent that the man was being taken seriously, he became angry and tried to calm the fears that Grodzinski had aroused.

However, Knijnikov was constrained by the cult of secrecy, which, despite *glasnost*, was still imposed upon those who worked under the aegis of the Ministry of Medium Machine Building. His data was classified, and he had no wish to be arrested by the KGB or lose his job. Still, the fact that Grodzinski's article had been published without dire consequences for the author or the editor of *Nature* suggested some political support for the position he had taken.

In the second half of 1988, Knijnikov wrote a lengthy and considered article on the subject that he had studied for so long. At last he dared to warn his readers of the dangers of a low collective dose of radiation, predicting that over the next sixty years there might be as many as twenty thousand additional deaths from cancer as a result of the Chernobyl disaster. When he submitted the article to *Pravda* it was turned down, and he was told that they had a surfeit of articles on this subject. If he had predicted, say, two million deaths, it might be have been of interest to *Pravda*'s readers: as it was, what he had written said nothing new.

A further blow to the official position came from the Soviet Union's most distinguished hematologist, Professor Andrei Vorobyov—the same Vorobyov who had served as the government's principal expert on the hazards of radiation on its own medical commission, and who, in the first days after accident, had told ministry officials that a dose of up to one hundred rems would not lead to leukemia. Now, in an article in *Novy Mir*, he warned his readers that even a dose of a single rem could have unfortunate consequences, and that in some zones the dose from Chernobyl had been ten times higher than the official figures suggested. Using a sophisticated process that included electronic paramagnetic resonance tests on tooth enamel, Vorobyov had discovered this in people living in the Gomel region of Belorussia. Moreover, the government had only given the figures for strontium and cesium, when there were many other radioisotopes that should be taken into consideration.

Government scientists retaliated, suggesting that political animosity colored Vorobyov's attack on Ilyn. The process of biological dosimetry used by the Institute of Biophysics was accepted as accurate all over the world. Vorobyov's method was too complex and expensive to be widely used, and the extrapolations made from a sample were inevitably open to doubt. While Vorobyov charged that Ilyn and his team were dishonest, deliberately lowering their estimates of the dose for political expediency, the scientists at the Institute of Biophysics and his former colleagues at Hospital No. 6 felt that Vorobyov's judgment was clouded by rancor and ambition. Certainly he had good reason to hate a regime under which his father had been shot and his mother imprisoned for eighteen years, but this should not lead him to feed false facts to the populace or democratic politicians or to turn against those who had helped him in the past.

3

It was too late for reasoned debate. In the course of the summer of 1988, to the consternation of its author, Mikhail Gorbachev, *glasnost* had taken on a life of its own. Treated at first with profound cynicism by the mass of the Soviet people as just another empty slogan, it had been gradually and cautiously tested and found to have more substance than was at first supposed. In the vanguard of those who hoped for a genuinely more open and democratic society was a group among the licensed intellectuals of the Writers' Union, and the primary issue that they used to test the limits was the critical condition of the environment.

This had several advantages. The subject had not been treated by either Marx or Lenin, and so had no counterrevolutionary connotations. It was also a palpable by-product of the era of stagnation—more vivid, in its effects, than falling production or inadequate investment. Moreover, it aroused real fear, particularly after Chernobyl, which became greater than citizens' habitual dread of the KGB. When it became apparent that people were more likely to suffer from the effects of radiation, pollution or pesticides in their own homes than from imprisonment in the gulags, a new resolve entered the hearts of those who until then had been cowed by the residual menace of the regime.

Nowhere was this more apparent than in those republics that had never willingly joined the Soviet Union. Lithuania was a good example; once an empire stretching from the Baltic to the Black Sea, and, like Latvia and Estonia, having more in common with Scandinavia than Muscovy, it had been the last nation in Europe to adopt Christianity, converting en masse in the fifteenth century from pagan tree worship to Roman Catholicism, not Russian Orthodoxy. Part of the Russian empire until 1918, it became independent in the wake of World War I, but was seized by Stalin in 1940 after Hitler had conceded it to the Soviet sphere of influence in the Molotov-Ribbentrop pact. Never resigned to its fate, anti-Soviet partisans had remained active for some years after the end of World War II. Even forty years later, at the time of Chernobyl, the nonindigenous population was smaller there than in any other non-Soviet republic, and the few Lithuanians who lived in other parts of the Soviet Union had mostly been deported there against their will.

The first test of *glasnost* in Lithuania came on 12 February 1988, when a small group held a rally to celebrate the Independence Day of the prewar state. This provoked an angry reaction from the Communist authorities; administrative sanctions were taken against many of the offenders, and a media campaign was mounted against this reactionary nostalgia for the "bourgeois" state.

It failed. Protected by their membership in the Writers' Union, Lithuanian intellectuals formed a number of unofficial clubs to discuss and reevaluate their recent history. A revisionist article appeared in a newspaper for young people; the author was vilified by the party, but the intellectuals took his side. Following news that a Popular Front had been formed in Estonia, five hundred members of the intelligentsia—authors, artists, scientists and academics—held a meeting at the Academy of Sciences on 3 June to launch a movement known as Sajudis to promote "openness, democracy and sovereignty."

Throughout that summer, Sajudis organized demonstrations to promote its agenda: democratic elections, access to the media, the national flag, but also the dangers of nuclear power. On 15 July about two thousand people protested in front of the Lithuanian Supreme Soviet against the construction of a third fifteen-hundred-megawatt RBMK reactor at the Ignalina nuclear power station, which had proved to be an issue of acute popular concern. Two months later, on 16 September, two hundred thousand people formed a human chain around the Ignalina plant to stop work on the third unit, and pickets were formed in different Lithuanian cities to call for an international investigation into the management of the power station.

The same explosive mix of ecological anxiety and patriotic feeling was evident in the Ukraine. Second only in size to the Russian Republic among the those that made up the Soviet Union, it had never received its due. Its black earth was celebrated for its fertility, it had developed a considerable industrial capacity and its population was as large as that of France, but it had failed to impress its identity on the outside world. Divided in the nineteenth century between the Russian and the Austrian empires, its large estates, as at Chernobyl, were often owned by Poles. By and large a peasant people whose language was not taught in schools or used by the educated classes, it nevertheless produced great poetry, like that of the liberated serf Taras Shevchenko.

In the Revolution of 1917, the Ukrainian peasantry rose against the Polish aristocracy, often slaughtering landowners and burning down their palaces and country houses. Ragged but brutal armies fought for the Reds and the Whites in the civil war. The soldiers of Hetman Petlyura and the anarchist Makhno indulged in atrocious pogroms of Jews, while Trotsky's troops massacred Pilsudski's Poles, the nationalist Ukrainians and anyone else they deemed a "bourgeois" or an enemy of the people. Between 1918 and 1920, the people of Kiev were governed by fifteen different regimes. The final triumph of the Bolsheviks was followed by the horrors of their collectivization of agriculture and the extermination and deportation of millions, succeeded by the death from starvation of many millions more.

No sooner had one vile episode in Ukrainian history come to an end than another started. In the 1930s came Stalin's terror, and when this ended, the Ukraine was invaded by the German army and, in its wake, the murderous *einsatzgruppen*, or action units. Ukrainian Jews were exterminated, some to be buried in a ravine known as Babi Yar on the outskirts of Kiev. So too were their gentile compatriots, accused of collaborating with the Soviet partisans. Twenty million Soviets were killed during World War II, a large proportion of them in Belorussia and the Ukraine.

Given these thirty years of unparalleled disruption and slaughter from 1915 to 1945, and the totalitarian tyranny imposed during the four decades that followed, it is hardly surprising that the Ukrainian people reacted cautiously to the promise of *glasnost* and *perestroika*. Earlier there had been a false dawn under Khrushchev, considered a liberal in the West, but to the Ukrainians Stalin's *gauleiter* in Kiev, noted for his zealous persecution of the Christian religion. What else should they expect from a Gorbachev or a Ryzhkov? Once bitten, twice shy.

Little by little, however, it became clear that while the old guard could still count on Yegor Ligachev and his group in the Central Committee, they were by no means all-powerful. The failure of Ligachev's move against Gorbachev in the spring of 1988 was followed by a formal reorganization of the Politburo at the end of September—Gorbachev's "coup" from above—in which Ligachev lost control of ideology and was given the poisoned chalice of agriculture. Indeed, the very department of ideology within the Central Committee was abolished, being replaced by a commission headed by Vadim Medvedev, at one time Yakovlev's deputy in the department of propaganda.

The earliest manifestation of any kind of opposition in the Ukraine came, as mentioned earlier, in a letter in *Literaturnaya Ukraina* soon after the accident at Chernobyl, protesting against the building of a nuclear power station at Chigirin. At the same time authors like Vladimir Gubarev, Yuri Shcherbak, and Volodomyr Yavorivsky began to write books blaming the disaster on the operators and the era of stagnation. In late 1986 and 1987, the debate about nuclear power continued in the pages of *Literaturnaya Ukraina,* and there were even conferences and seminars on the question, but no criticism of the government's program was allowed to appear in the mass media.

In December 1987 a group from the writers' and the cinematographers' unions founded a club in Kiev to discuss environmental issues. They called it Zelenij Svit, or Green World. Among the thirty or so founding members was Yuri Shcherbak. In March 1988, at a conference held by the Writers' Union, the club was expanded into a pan-Ukrainian ecological association with opposition to nuclear power as the main plank of its program. As it became apparent that these pioneers had gotten away with this form of protest—indeed, that it was encouraged at the highest levels in the party—thousands of others joined. In July 1988 twenty thousand signed a petition against the continuing construction of a nuclear power station in the Crimea. The membership of Green World grew from thirty to five hundred thousand, and in the spring of 1989 Yuri Shcherbak, running against the officially sponsored candidate of the Communist party, was elected on a Green World ticket to the Congress of People's Deputies in Moscow.

Even before Shcherbak's election, the government in Moscow had slowly but inexorably given in to popular pressure to cut back its program for nuclear power. As early as 1987, plans for further nuclear power stations in Kiev and Odessa had been abandoned. In January 1988 construction was halted at the Krasnodar station, and that autumn it was announced that a planned nuclear power station in Minsk would be converted to fossil fuel, while reactors in Armenia and Azerbaijan were closed.

In the spring of 1989, when the elections had established decisively the public's ecological anxieties, expansion was halted at the South Ukraine, Kursk, Smolensk, Khmelnitsky and Rovno nuclear power stations. The power station in the Crimea was converted into a training center—Velikhov had agreed that the site was susceptible to earthquakes—and a decision was made to convert the Chigirin power station to fossil fuel. In the autumn of 1989, Green World organized mass demonstrations in Kiev against nuclear power, and in the spring of 1990, plans for the expansion or construction of nuclear power stations were abandoned at Gorky, Archangelsk, Karelia, Ivanova, Tataria and Rostov.

Lastly, on 1 June 1989, the Supreme Soviet of the Ukraine decided to decommission, at some unspecified future date, the three working reactors at Chernobyl.

4

Among other successful candidates on the democratic ticket in the election to the Congress of People's Deputies in 1989, the first since the Revolution with a measure of genuine democracy, was Professor Andrei Vorobyov, Academician Yevgeni Velikhov and the author who had made his name because of Chernobyl, Volodomyr Yavorivsky. His novel, *The Star Called Wormwood*, which had so bitterly offended those who had worked at the power station, had been a great popular success. Its portrayal of slovenliness and corruption among the personnel satisfied not only the party line but also the public's long-suppressed appetite for controversy and sensation. Although written in complete conformity with the government's new policies of *glasnost* and *perestroika*, it had the paradoxical effect of earning a reputation for its author as a radical critic of the Soviet system.

If the book altered Yavorivsky's public image, it also changed the inner man. The son of peasants from around Vinnitsa in the southwest Ukraine, he had been shaken by what he had seen while interviewing the operators and liquidators in the thirty-kilometer zone for his book. The need for change became more than a policy of the Twenty-seventh Party Congress; it became a cause to which Yavorivsky would devote his life. When RUKH was formed—the Association for the Promotion of Perestroika in the Ukraine—he became a member. A talented orator, he delivered an inspiring speech to the first congress of RUKH, looking back over Ukrainian history, then asking, "What sort of people are we? What kind of people have we become?"

Earlier than many latter-day nationalists, Yavorivsky resigned from the Communist party and abandoned his career as a writer to become a full-time politician. In appearance he was handsome, even romantic, and his second wife was an actress. Some who sympathized with his objectives saw Yavorivsky as a long-winded poseur, but these qualities have never impeded the career of a politician, and they did not prevent his triumph. Running on the platform of RUKH, he was elected to the Congress of People's Deputies in 1989, and a year later to the Ukrainian Supreme Soviet in Kiev.

Once elected to the republican assembly, Yavorivsky set to work to frame a "Chernobyl law" that would provide aid and compensation to the victims of the accident. To accomplish this he allied himself with the president of the Chernobyl Union, Volodomyr Shovkovshytny.

Like Yavorivsky, Shovkovshytny was the son of peasants from a village south of Kiev. Nominally members of a state-owned farm in which five hundred people worked three thousand hectares of land, his parents were largely self-sufficient, with their own painted wooden cottage, fruit trees, eight pigs, eighteen geese, thirty hens and a private plot of land. They had few happy memories of the Communist era. His mother's family had been forced to cut down their fruit trees because they made them liable for a tax they could not pay. His father's family, which had owned a horse, a cow and a pig, had been denounced as kulaks, thrown off their land and evicted from their home. His maternal grandfather had been sentenced to ten years in a gulag for resisting collectivization; his father's mother had died in the famine. Two great uncles had died in the war and a third had returned from a German camp only to die in one of Stalin's gulags in Siberia.

Volodomyr had therefore been raised with a secret loathing of the regime. Although his parents were upright and god-fearing, their passion was for a Ukrainian nation, and the small shrine in their living room was to their national poet, Taras Shevchenko. Young Volodomyr had inherited this love for Shevchenko's verse, and after reading all the books in the village library, dreamed of becoming a writer himself. It remained a dream. After serving in the army, studying geology in Kiev, and prospecting for uranium in Siberia, he took a job at the Chernobyl nuclear power station, where his brother was already employed as a dosimetrist.

Never a high-flying engineer, Volodomyr Shovkovshytny worked first in the chemical workshop, and later for Nikolai Steinberg in the turbine hall. His principal interests were outside his work; he liked sports and, above all, literature. He wrote light verse and humorous stories, which were published in Pripyat, played the guitar, staged plays in the House of Culture and got to know other aspiring writers like Lubov Kovalevskaya.

At the time of the accident, Shovkovshytny had been away in Moscow taking exams to obtain higher qualifications. He returned to Pripyat only to be evacuated with his wife and children. That autumn, he resumed work at the power station but he still wanted to be a writer and a year later he was offered a place at the Literary Institute in Moscow.

The two years Shovkovshytny spent in Moscow opened his mind. He had an opportunity to study the great works of Russian and Ukrainian literature, to see the contemporary plays and to listen to the new political ideas circulating in the capital city, where *glasnost* had advanced further than it had in Kiev. Just as his dream of becoming a writer had now come true, so the dream of a Ukraine that was free of both Communism and the tyranny of the Kremlin had suddenly ceased to be absurd. Certainly the ecological question was still the subject of the most urgent agitation, but to Shovkovshytny green remained a mix of blue and yellow, the traditional colors of the Ukraine.

Like Yavorivsky, Shovkovshytny was a dashing figure, with a drooping mustache and unquestionable charm, and he soon gained a reputation in Kiev both for his personal integrity and his political ideas. In Moscow at the time of the elections for the Congress of People's Deputies, he was asked by his friends in the Chernobyl Union to run for election to the Ukrainian Supreme Soviet a year later. It was a long shot; the party still had great influence, and on the list of his opponents was a deputy minister, a first secretary and the mayor of the residential district on the east bank of the Dnieper where he lived. But many of the former residents of Pripyat had been rehoused in this suburb, among them one thousand members of the Chernobyl Union, and on the second round of voting, Shovkovshytny won the seat.

5

In the campaign for these elections to the republican Supreme Soviet, environmental, scientific and political issues had all become inextricably linked together in the public mind. No longer was Chernobyl the product of Brezhnev's era of stagnation—the apotheosis, as Legasov had said, of the incompetent running of the economy. Now it was the product of the Communist system itself, the apotheosis of Bolshevik duplicity and indifference to the welfare of the people. But to prove their point, the democrats had to establish that there had indeed been duplicity and indifference: that the children of Pripyat had deliberately been left to play in air laden with radioactive iodine 131; that the citizens of Kiev had been made to march up the Khreshchatyk on 1 May while the party bosses had sent their families out of the city; and that to save the state a few rubles, millions of people were still living on contaminated land.

The more vigorously this line was pursued by the enthusiastic activists of Green World, the more alarmed became those still living in the contaminated areas. On 30 March 1989, a meeting was held to discuss the situation in the House of Culture in Narodici—ironically, one of the towns to which the inhabitants of Pripyat had first been evacuated. People came from the surrounding countryside; the hall was soon full, and loudspeakers were placed in the corridors outside to enable latecomers to hear what was said.

First the chairman of the local civil defense committee declared that on the day after the accident the radiation levels in Narodici itself had measured three roentgens per hour; yet no one had been told and nothing had been done to protect its citizens. A dosimetric expert from Romanenko's Center of Radiation Medicine who had come to the meeting protested that this was impossible, but he was contradicted by a member of the public who said he had data from the Center itself that revealed that the level of cesium in the soil just north of Narodici was more than fifty curies per square kilometer. "How can children be expected to live in such conditions?" he asked.

Not merely in Narodici, but also in nearby Polesskoe, people demanded to be evacuated from the contaminated zones. Claim and counterclaim, made by the government on one side and the Green activists on the other, drove the often unsophisticated residents into a state of hysteria. "I am still young, I want to live," cried a woman at one of the village meetings, "and I want my children to live. But when I give my children a glass of milk, I feel treacherous, because I do not know how much cesium it contains. I have already been in the hospital three times, and so have my children. Yet before the Chernobyl tragedy, we were all healthy."

Even in lightly contaminated areas, people demanded the extra monthly allowance to buy "clean" food. Yet when the government agreed to their demands and paid this "coffin money," it was taken as proof that they had indeed been living in danger all along. Most used the money to buy vodka, the traditional antidote for both radiation and despair.

Against this barrage of invective, Ilyn and Knijnikov were obliged to amplify the official line. In April 1989 Ilyn wrote an article published in both Ukrainian and Belorussian periodicals stating that in the 786 zones affected by the accident in Chernobyl, the average dose received by people was six rems. Blood tests had shown that twenty-six hundred had received a higher dose of up to 17.3 rems, and eight hundred of these a dose that was higher still. Ilyn predicted that over the next thirty years there would be an additional thirty cases of thyroid cancer, and an additional sixty-four deaths from leukemia and other cancers. In an interview, Knijnikov conceded that "there were doses of radiation rather high for the thyroid gland" and that "theoretically we can expect a higher cancer incidence in medical personnel. There is a risk, though it is not high, about one percent, but its negative effects can be reduced by timely diagnostics and adequate treatment." In the same article, Vladimir Asmolov, the head of the nuclear-safety department at the Kurchatov Institute, referred to the Soviet Greens as "the most ignorant Greens in the world."

This new candor did nothing to calm the agitation, or to save Ilyn and Israel from the charges that they were responsible for covering up a heinous crime. The campaign against them was led by the radically pro-*glasnost* newspaper *Moscow News*. In February 1989 it published an article by Vladimir Kolinko, a journalist who had covered the Chernobyl accident for the Novosti News Agency, describing the grotesque livestock born in Narodici and claiming a dramatic increase in thyroid disease and a doubling in the incidence of lip and mouth cancer since the accident. He also wrote that there had been an increase in the incidence of infant mortality, and that abortions had been recommended to women in the zone. Kolinko's article was reproduced in a Ukrainian newspaper for young people, *Molod Ukrainy,* which ensured that his alarming statistics reached those concerned.

Kolinko's claims were quickly refuted in *Pravda Ukrainy* by scientists from Romanenko's Center of Radiation Medicine in Kiev. The head of the epidemiological laboratory wrote that because of the migration of the population from the Narodici region after Chernobyl, the number of patients treated for cancer had actually *dropped.* There were only three cases of lip cancer in 1987 and 1988, whereas there had been seven in an earlier two-year span. The director of the Center's Institute for Clinical Radiology showed that the birth and death rate of babies had remained stable in the period, and that while local doctors might have recommended abortions, this had never been sanctioned by the authorities in Kiev.

The claimed genetic mutations in livestock were also dismissed with a certain measure of scorn. The Petrovsky collective, where Kolinko had claimed there were sixty-two freak calves, had been found by experts to have produced eight; this amounted to about 2 percent of live births, the same number as had been recorded in three farms in the Polessia region on a veterinary inspection before the accident. These genetic anomalies were thought to have come either from inbreeding or from an increase in nitrates and an absence of microelements in the soil.

Acknowledging that scare stories of this kind had been made possible by the obsessive secretiveness of the authorities in the past, the editor of *Pravda Ukrainy* sought to add authority to the official point of view by seeking the opinion of the American expert on radioactive contamination, Professor Richard Wilson of Harvard University. Wilson condemned Kolinko's article as an attempt by the Soviet press to ape the kind of sensationalism so common in the West.

In a further attempt to reassure the public, the Ukrainian Council of Ministers published for the first time a detailed account of the consequences of the accident in the newspaper *Radyanska Ukraina,* including maps of the contaminated zones. Intended to reassure the population, in point of fact, this official candor revealed areas of contamination previously unknown to the general public, and higher levels of contamination in meat, fish and mushrooms than had been hitherto supposed.

Nor did it stop the search for scapegoats by *Moscow News*. "Thousands of people are asking today: who is responsible for the favorable radiation readings in many Belorussian and Ukrainian villages whose inhabitants now have to be moved to new places?" In its April 1989 issue it described the "Chernobyl syndrome" by the formula that "the damage caused by the mistake is directly proportional to the length of time it is hushed up." In May, it posed the question put by the Belorussian critic and filmmaker, Ales Adamovich: "Who should answer for such an unforgivable three-year-long disregard for the health and interests of tens of thousands of people?"

6

Passions ran high, and some of those who had held power at the time of the accident broke ranks to shift the blame. On 12 July, at a televised session, the Supreme Soviet was asked to ratify the reappointment of Yuri Israel as chairman of the State Committee of Hydrometeorology. He immediately ran into a barrage of questions about his role in the aftermath of the accident. Some speakers criticized him, others defended him, and then came a dramatic intervention. Valentina Shevchenko, the proletarian virago who had stood next next to the Ukrainian party leader Shcherbitsky to watch the May Day parade on the Khreshchatyk in Kiev in 1986, walked up to the podium to speak.

"I am one of the representatives of the political leadership of the Ukraine that we are talking about today," she said. "In addition, the Chernobyl zone is today in my constituency. I would . . . simply like to read out to you some excerpts from a document." She then read from the document that the Ukrainian leaders had insisted should be signed by Ilyn and Israel on 7 May, stating that the levels of radiation in and around Kiev were safe. Then she turned to Israel. "Surely you remember, Yuri Antoniyevich, when we invited you to the Politburo of the Communist Party of the Ukraine Central Committee? Do you remember how you talked to the whole Politburo, and how it was proposed that you should write a note with recommendations and conclusions, and how you and Comrade Ilyn worked the whole day on this note, and the Politburo of the Communist Party of the Ukraine Central Committee held a session at half-past eleven at night to examine your conclusions and recommendations? You will also remember, then, that when you were sitting at the table opposite me, I asked you a question: 'Yuri Antoniyevich, what would *you* do if your own grandchildren were in Kiev?' You remained silent. And on the basis of your silence, we insisted that the Politburo of the Central Committee of the Communist Party of the Ukraine adopt a resolution to evacuate children from the city of Kiev. I am grateful to all the fraternal Union republics that took in our children; as a result of this we were able perhaps to prevent more widespread illness among our children."

Then Shevchenko took up the question of the evacuation of Pripyat. The Ukrainian leadership, she said, had been ready to evacuate at 8:00 A.M. on 28 April, but the decision had only been made by Scherbina that afternoon. He had been the head of the commission, and all the decisions he had made had been carried out by the government of the Ukraine "working most efficiently round the clock." She conceded the complexity of the task they had faced, but "today we still do not know what to expect, and I think that Yuri Antoniyevich, occupying such a high state post and being responsible for exceptionally important work, should occupy a standpoint not of compromise, but . . . of great principle. . . . I will vote against you, Yuri Antoniyevich, and appeal to the deputies from Ukraine to vote against you."

This intervention brought the Soviet prime minister, Ryzhkov, to the rostrum. "I should like to throw light on certain aspects of one problem about which Comrade Shevchenko spoke so fervently just now. This is the issue of the Chernobyl disaster. We must quite soberly evaluate the situation there at the time, and what is happening there today. I asked to speak because exactly three days after the accident an operation group of the Politburo was given broad powers to eliminate the extremely complex situation that had developed at that time. There is no need for me to describe in detail what we did. We did everything that we could have done. Everything that we understood, we did to save people. This was true, I think, over the two years that the commission worked from

29 April 1986 to 1988. In 1988, the commission did indeed cease to exist, and its functions were handed over to a government commission. Let me say, however, that at that time we faced a most difficult situation. It is easy enough to argue today about why such and such was not done in such and such a way, and so on. But at the time every moment counted and decisions had to be made quickly, responsibly and in a professional way, because it was a matter of people's lives.

"We heard here of a meeting of the Politburo of the Ukraine on 7 May. I don't know about that. I was not present. But I know that I was with you on 2 May. Do you remember? Do you? Do you remember that we went to the zone with the political leadership? And that was before 7 May. I do not want to accuse anyone, but I think it is quite wrong to level accusations today against those people who did everything they could to save the situation.

"A second matter. It was a special situation, and I have to say that enormous human effort and material resources were required to put out the fire. We did not know what to expect. The consequences, could have been far worse than those that did occur. We involved our finest scientists. They were physicists—among them the late Valeri Alekseyvich Legasov, now dead. He helped us with his knowledge as a physicist. He made recommendations to the operational group. And I have to say that he does not deserve to be forgotten. Then there was the physicist Velikhov—I cannot see him here today—who was also there, who also gave advice. Or take Ilyn, Academician Ilyn, who made certain decisions on health care. And I would like to say that among the scientists whom we gathered there at the disaster site, the very best of the scientific world, was Comrade Israel whom we are discussing today.

"I am convinced that if we had not attracted at that time the very cream of our society—our scientists—we could not have coped with that tragedy, or even if we had, then we would have made even more mistakes than, unfortunately, are apparent now. Today we should be thanking the scientists who helped so much to eliminate this national tragedy."

XII

1

Ryzhkov's eloquent speech, an unprecedented outburst of emotion from such a reserved man, did not stop the dissident clamor. Ryzhkov himself, apparently so decent and reasonable, the very best that the system could provide, was nevertheless a product of the system, a party boss, and damned as such by the new nationalists and democrats in Belorussia and the Ukraine.

Nor did *Moscow News* let up on its campaign to punish those responsible for Chernobyl. In October 1989, in an article entitled "The Big Lie," it printed the transcript of a round-table discussion on the cover-up held by Yuri Shcherbak and Ales Adamovich, both now people's deputies; Vladimir Kolinko, the journalist from the Novosti News Agency; Valentin Budko, the first secretary of the Narodici District Communist Party Committee; and two other people's deputies from the affected regions, Yuri Voronezhtsev from Gomel and Alla Yaroshinskaya from Zhitomir. The latter described the anguish suffered by those still living in the contaminated zone. "I heard a woman addressing a meeting. She said: I come home and my little son stands near a cup of milk. On seeing me, he gets frightened that I'll scold him because he isn't supposed to drink the milk, so he tells me, Mommy, don't swear at me, I've only dipped a finger into the milk . . ."

Who was responsible for this cruel state of affairs? "The lying started three and a half years ago," said Shcherbak, "and I believe that we still do not know the most dramatic truth about the accident." There must now be a thorough investigation to discover "who made the decision not to notify the public about [this] global catastrophe." Adamovich wanted to punish those responsible for "the crime that started in 1986, continued and is continuing. . . . People who are guilty of these crimes, all these lies and frauds, of concealing the truth . . . will not be able to change the situation. . . . To conceal their lies, they will have to continue prevaricating and lying. Therefore those who have not yet managed to retire on pension must quit their posts."

The round-table participants named names. As the head of the commission, Boris Scherbina had signed the order that classified all information about radioactive contamination in the thirty-kilometer zone. But Vladimir Marin, now chief of the Nuclear Power Division of the Bureau of Fuel and Energy, but then responsible for the nuclear power department of the Central Committee, had known of the accident an hour or so after it had happened. There were also the physicists and scientists. "How could doctors sign documents hiding the truth from the people, thereby dooming them to suffering?" asked Adamovich. "Our science and medicine," answered Shcherbak, "have turned into servants of the political system. This is the most horrible thing that can happen to science."

Most clearly culpable in this respect were Professor Yuri Israel and Academician Leonid Ilyn. The latter had not only corroborated "the big lie" but was also responsible for establishing the permissable dose of thirty-five rems which Velikhov had told Adamovich "came out of the blue." Admittedly the World Health Organization had approved the thirty-five-rem norm after a visit to the affected area in 1989, but when the French atomic specialist Professor Francis Pellerin had been asked why he supported it, he had said that the Soviet Union lacked the resources to make it any lower.

A particular target for Shcherbak was Ilyn's ally in Kiev, the minister of health, Anatoli Romanenko. "In the Ukraine today, at literally every meeting, people demand that Anatoli Romanenko, minister of public health of the Ukrainian SSR, should be called to account." Although in the United States at the time of both the accident and the May Day parade, he was now held responsible for the delay in alerting the people of Kiev to the dangers of radiation, and in ascribing the suffering of those living in the contaminated territories to "radiophobia." He had lost the protection of the long-serving party leader Vladimir Shcherbitsky, who had retired in September; the last of Brezhnev's satraps, and a consistent opponent of both Gorbachev and *glasnost,* Shcherbitsky astounded everyone by surviving for so long. In November Romanenko followed. His resignation came when the Politburo of the Ukrainian Communist party met to discuss Chernobyl. The reason given was the burden imposed by his dual role as minister of health and director of the Center of Radiation Medicine; he retained the second position, which, as a national appointment, was the gift of Ilyn, not of the local party leaders. He was replaced by his young deputy, Yuri Spizhenko, until then a supporter of his superior's stand.

At the meeting of the Politburo, the first secretary of the Kiev Regional Committee, Grigori Revenko, who had been deeply involved in the Chernobyl crisis from the beginning, blamed both RUKH and Green World for fomenting fear among the population by spreading unscientific rumors. However, by February of the following year, Spizhenko admitted to the Ukrainian Supreme Soviet that there was indeed a medical crisis in the controlled regions; the thyroid glands of fifty-eight hundred children and seven thousand adults had been adversely affected by radiation, and a large number of the two hundred thousand liquidators living in the Ukraine required careful medical attention. He blamed the past failure of his ministry in appreciating the magnitude of the crisis on the classification of the data by the "so-called Third Division of the USSR Ministry of Health."

2

Although the loudest protests over Chernobyl came from the Kiev region, only 20 percent of the radionuclides spewed out of the fourth reactor had in fact fallen on the Ukraine. Ten percent had settled on parts of the Russian Republic, and the remaining 70 percent had fallen on Belorussia.

The initial docility of this Soviet republic in the aftermath of the accident sprang partly from the nature of its people and partly from its particular history. In a land largely composed of forests, meadows and swamps, its inhabitants were often dismissed by the supposedly more sophisticated Muscovites and Ukrainians as *boulbash,* "onion eaters." Living in isolated villages, the peasant population had, even at the best of times, little contact with the outside world. When it did impinge on them, it had only brought suffering: World War I, the Revolution, the collectivization of farms, and finally World War II, in which a quarter of the population had been killed. Even before then many of the ablest Belorussians had either emigrated or had been exterminated. One of the by-products of *glasnost* was the discovery, in the woods at Kuropaty on the outskirts of Minsk, of the remains of three hundred thousand people shot under Stalin by the NKVD.

At the time of the accident, the general secretary of the Belorussian Communist party was N. Slyunkov. A tough, conservative industrial administrator with a penchant for grandiose projects, he combined contempt for intellectuals with a determination to keep the Kremlin off his back by demonstrating unquestioning loyalty to the Politburo.

Immediately after Chernobyl a Belorussian nuclear physicist had gone to the Central Committee in Minsk to warn of the dangers from fallout, but was only taken seriously when he started to read off measurements in the offices of the Central Committee itself. However, the levels were low; only in the provinces around Mogilev and Gomel were they high, far from the children and grandchildren of the party leaders in Minsk. Consequently there was none of the anxiety in the Belorussian Politburo that was so apparent among the Ukrainian leaders in Kiev.

Moreover, because Minsk was a smaller city, there were fewer indigenous scientists and intellectuals to question the party line. At the Institute of Nuclear Energetics, a professor of nuclear electronics, Stanislas Shushkievicz, noticed the increase in the levels of radiation and telephoned the authorities in alarm, only to be told that everything was under control. Later his dosimetric equipment was impounded, but he secretly designed and made new instruments. In addition, students at the university were taught how to measure radiation using samples of condensed milk from the Gomel region.

Evacuation had been organized from the Belorussian sector of the thirty-kilometer zone by the area civil defense chief, headquartered in Khoyniki. Twenty-six thousand people and thirty-six thousand head of cattle were removed from contaminated land. In Minsk, a deputy chairman of the Belorussian Council of Ministers headed a committee for the elimination of the consequences of the accident. The Belorussian health minister, Nikolai Savchenko, acknowledging that initially "many economic leaders and citizens displayed unconcern and elementary medical incompetence," coordinated the efforts of the medical commission in Moscow. By June 1986 sixty thousand children had been sent to holiday camps in a belated attempt to remedy the failure to provide iodine prophylactics in time.

Despite the heavy-handed regime of General Secretary Slyunkov, some protests about inadequate care were possible. In December 1986 the Moscow magazine *Argumenti i Fakty* published the complaints of four women living in the towns of Bragin and Komarin to the Belorussian Council of Ministers that while local officials had evacuated their own children, no one had bothered about those of the ordinary people. The same suspicion of their local leaders was felt by workers in the October Revolution collective forty kilometers north of the Chernobyl nuclear power station. Alexei and Antonia Dashuk and Fyodor and Olga Tithonenko, all about sixty years old, were told that it was safe to stay in their homes, but they noticed that the inhabitants of the two neighboring villages were being evacuated and that these happened to be the villages in which the collective bosses lived.

With the departure of the collective, the Dashuks and Tithonenkos were told to join the local state-owned farm at Strelichevko. The Dashuks continued to live in their own home, a spacious wooden house with five rooms, a veranda, a cellar, a barn and half a hectare of land. They were told not to go into the forest, but the state farm continued to produce food, which it sold to the state. They grew and ate their own vegetables and made their own alcohol, but the cows were gone, and they were given an allowance of thirty rubles each to buy milk imported from outside the region.

3

Despite the liquidation of the Belorussian intelligentsia by first the Bolsheviks and then the Nazis, there were some survivors whose children took advantage of *glasnost* to promote cautiously the interests of their country. Zyanon Poznyak, an archaeologist, the grandson of an early nationalist leader, began the excavations at Kuropaty that uncovered the remains of the Bolsheviks' victims and, following the example of the Baltic states, founded a national front. Stanislas Shuskievicz, the scientist at the Institute of Nuclear Energetics, whose father had been sent to the gulag by Stalin for writing children's stories in the Belorussian language, ran for the Supreme Soviet and was elected a people's deputy. However, the man who was most outspoken in his criticism of the way his compatriots were treated after Chernobyl was not the child of an intellectual, but a man who had been born in a village in central Belorussia and had started fighting the Germans at the age of fourteen.

Even before the accident, Ales Adamovich had been a critic of nuclear power. Never considered ideologically sound, he had nonetheless established himself as a writer, critic and filmmaker, expounding a form of humanism that diverged from the party line. With the advent of *glasnost*, he had been among the first to insist on naming Stalin as the perpetrator of some of history's most atrocious crimes, accusing him of deliberately organizing the famine of 1930–31 and referring to him and his henchmen as "butchers." To orthodox Communists this was blasphemy, and in September 1988, outraged by the failure of the authorities to punish Adamovich for this slander, a retired public prosecutor brought a charge of defamation against him in the courts. It was dismissed.

In the same month, there was a further triumph for Adamovich when it was announced that the nuclear power station being constructed near Minsk was to be converted to fossil fuel. His reputation rose both in his native Belorussia and among his colleagues in Moscow. After the failure of Ligachev's coup and the triumph of Yakovlev's liberal interpretation of *glasnost*, it was the Cinematographers' Union that had first swung behind the new party line, and it was by the Cinematographers' Union, not the Belorussian people, that Adamovich was sent as a delegate to the Congress of People's Deputies in March 1989.

To Adamovich, the campaign against Stalinism and nuclear power were one and the same; Chernobyl and Kuropaty were both aspects of the same historical phenomenon. The cruelty that had led the Bolsheviks to liquidate Belorussian patriots or to starve the Ukrainian kulaks in the 1930s was the equivalent of their successors' condemning the Belorussian people to a lingering death from the effects of radiation and hiding the truth of what they had done with the same lies and false propaganda. Like Shcherbak in the Ukraine, Adamovich was determined to expose those responsible for "the big lie" and to remove them from power.

Just as in neighboring Lithuania, where mass protests against the regime had begun by demonstrating against the dangers of the Ignalina nuclear power station, so in Belorussia outrage over Chernobyl was at the forefront of the democratic campaign. A movement called Chernobyl Shlyaka—Chernobyl Way—organized demonstrations in the streets of Minsk with banners advertising the levels of contamination in the different areas of the republic. In the wide boulevards of the somber, neoclassical city, people in the milling crowd exchanged stories about illness in their region, while the democratic candidates running for election promised to fight for compensation and a Chernobyl law.

Once elected, the new deputies continued their agitation. The president of the Belorussian Academy of Sciences, Vladimir Plotonov, expressed doubts about the safety of the thirty-five-rem threshold. He was a mathematician and admitted his ignorance of nuclear science, but the remarks of the French scientist Francis Pellerin seemed to suggest that the Soviet government had left people in hazardous areas simply to save money. Of course there were experts in Belorussia. One, Professor Konoplia, like Grodzinski in Kiev, arguing that no dose of radiation could be considered safe, became a national hero.

Another, Professor Vladimir Matuchin, a Russian who directed the Institute of Radiological Medicine in Minsk, supported the government's point of view and, together with ninety-one other scientists on the State Commission for Radiation Safety, signed a letter supporting the thirty-five-rem control limit and the measures that had been taken for the protection of the population. Some months later, visiting Ilyn at the Institute of Biophysics in Moscow, Matuchin broke down in tears. He had been harassed and threatened by anonymous callers who said that if he continued to publish data in defense of the government, his family would be killed. Victor Knijnikov, the head of the laboratory at the Institute of Biophysics who had laid down the criteria for acceptable levels of contamination in food, was summoned to Minsk to give evidence at a session of the Ukrainian Supreme Soviet along with two of his colleagues, one of whom had prepared a paper describing how no notable changes in the health of children in the contaminated zones had been found. But after the threatening speeches of the Belorussian deputies, the paper's author declined to take the floor, leaving it to Knijnikov, a Jew, to face the charge of genocide.

4

The campaign against the government's experts on radiation was not confined to Minsk. In Leningrad, the director of the Institute of Radiological Hygiene, Professor P. V. Ramzayev, one of the foremost experts in Russia on the question, wrote an article that undertook an objective analysis of the radioactive danger in the affected zones. After its publication, he was told by an anonymous caller that if he continued he would end up with a hole in his head. Meeting Ilyn in Vienna, he said he would write no more on the subject of Chernobyl. "Leonid Andreevich," he said, "I have a wife and children. I must think of them."

In Kiev, the director of the Institute of General and Communal Hygiene, Academician Mikhail Shandala, had in the wake of the accident led a team into the contaminated zone to do exhaustive research on the nutrition of the inhabitants and had found it dangerously deficient. However, when a proposal was put forward by Belorussian politicians to resettle up to a million people from land he considered safe, he had added his name to the ninety-one others from the State Commission for Radiation Safety.

After Romanenko's removal as minister of health in the Ukraine, Shandala was quietly advised that he too should go. He not only had signed the letter but had given interviews "simplifying a complex situation." "You are a clever man, Professor," he was told, "but the tide of public opinion is running against you." Afraid not just for his professional position but for the safety of himself and his wife, Shandala moved to Moscow, where he was nominated for the post of director of the Institute of Preventive Toxicology.

To the democrats and nationalists of RUKH and Green World, this was all to the good. Shandala, Matuchin, Ramzayev, Israel, Ilyn, Romanenko and Knijnikov were the heirs of Beria's evil empire, which would only end when they fell from power. Their standing as scientists was no protection. Ilyn and Israel were threatened with prosecution, and now lived in fear of arrest and imprisonment. To have been in any way associated with the military-industrial complex, to have worked for an institute that was part of the Ministry of Medium Machine Building or the Third Division of the Ministry of Health or to have held notional rank in the KGB meant that they had collaborated with the crimes of the Stalinist state.

Just as the scientific issues were complex and open to different interpretations, so were the motives of many of those who joined battle in the aftermath of Chernobyl. On one side were the Soviet patriots— people like Ilyn, Romanenko and Guskova, who, unlike Alexandrov and Dollezhal, had no memories of life before the Revolution. Communism was the religion of Russia, and to love one's country was to uphold its faith. With direct lines to the Politburo, there was no divergence in outlook between these men and their leaders; to Gorbachev and Ryzhkov, as well, Mother Russia was the Soviet state.

Behind these commanders came the troops, the scientists like Knijnikov and the doctors like Baranov. Inevitably, given the nature of their expertise, they had been employed by the defensive and coercive organs of the state, one which itself had a military structure and tradition going back long before the Bolshevik Revolution. To the democrats, nationalists and Greens, the subordination of the scientists to their superiors and the subordination of their superiors to the ideological imperatives of the state were proof enough that the scientists could not be trusted, and the wretched populace of the controlled zones was easily persuaded to agree.

However, the legacy of Stalin and Beria had also affected their victims. There was the simmering loathing of those like Professor Andrei Vorobyov, Lubov Kovalevskaya and Volodomyr Shovkovshytny, whose parents or grandparents had suffered from the Bolshevik terror, and also the self-disgust of those who in order to get a good job or to see their books published had gone along with the party line. Among the doctors who most vigorously condemned Ilyn was one who had watched without protest when dissidents were committed as lunatics in the institute where he worked. There had not been many such inmates—few had dared to dissent before *glasnost*—but when it had become apparent in 1988 that criticism would not only go unpunished but was actually encouraged and might well be a good qualification for the future, many seized upon the issue of Chernobyl to establish their democratic credentials.

In this time of rapid historical transition, there were also men and women who bestrode the two worlds. At the time of the accident at Chernobyl, Yuri Shcherbak, the leader of Green World in the Ukraine, had been a figure of sufficient standing in the party to be a member of the Writers' Union. He had even formed part of the entourage surrounding Armand Hammer and Dr. Gale on their visit to Kiev and Chernobyl in the summer of 1986. Although Gale's later account of his two visits made no mention of him—the American named Romanenko and Shandala as his hosts—Shcherbak nonetheless recorded several conversations with this by now illustrious American, which he later included in his book.

Still referring to Lenin as "Our Leader," Shcherbak nonetheless revealed a perplexed respect for the American doctor in his clogs and blazer, and a certain awe at the luxurious fittings of Hammer's private Boeing; he also made a public confession of his cowardice in the pre-*glasnost* era:

> For long years before April 1986 I had been pursued by a feeling of guilt . . . because I, a native of Kiev, a writer, a doctor, had passed by on the other side of the tragedy of my native town . . . which had occurred at the beginning of the sixties: the damp sand and water accumulated in Babi Yar, which the authorities wanted to make into a recreation area, broke through a dike and poured into Kurenivka, causing . . . destruction and . . . death. . . . And why did I remain silent? I could have collected facts, the testimony of witnesses, I could have found out and named those guilty of this calamity. . . . But I didn't.

Chernobyl offered Shcherbak the chance to make amends.

5

In contrast to Shcherbak was Volodomyr Shovkovshytny, the president of the Chernobyl Union, who before the accident had been a man of no standing at all. Only moderately talented as a poet and unexceptional as a nuclear technician, he would undoubtedly have remained, like the Ukraine, in undistinguished obscurity had not Chernobyl reignited his national pride. Raised in a house with a shrine not to Marx or Christ but to his people's patriotic poet, Taras Shevchenko, it was not so much radiophobia as Russophobia that led Shovkovshytny to champion Chernobyl's victims. The "Czars and princelings" who been the objects of Shevchenko's loathing were now, for him, the party bosses and Soviet officials who had been indifferent at the time of Chernobyl to the fate of the ordinary Ukrainian people:

> So fly, my fledgling falcons, fly
> To far Ukraine, my lads—
> At least, if there you hardship find,
> 'Twon't be in foreign lands.
> Good-hearted folks will rally 'round
> And they won't let you die.

"Good-hearted folks will rally 'round / And they won't let you die" was the principle behind the Chernobyl Union, which from its inception in 1987 sought both to help and to represent the interests of those who had suffered as a result of the accident. In 1990, a second organization, Chernobyl Help, was set up in Moscow by Robert Tilles, the engineer who had constructed the "biological wall" in front of the ruined reactor. By means of telethons and other charitable drives, millions of rubles were raised to help the victims. Appeals were made abroad for medicine, as well as for help for the children of Chernobyl. Groups of children were sent on holidays abroad, to western Europe, Australia, and Cuba. Haunting pictures of children suffering from leukemia brought a generous response, particularly from West Germany; the Bavarian Red Cross sent large consignments of medical aid, which was then distributed by the Chernobyl Union.

Both these charities had their critics. Some, like Professor Vorobyov, argued that when children had suffered from radiation, it hardly helped them to increase their dose by sending them on long flights to sunny countries. Others felt that a taste of life in the West only lowered the children's morale when they returned to their villages in the Ukraine or Belorussia. The Chernobyl Union complained that Chernobyl Help hoarded its funds, keeping ninety million rubles on deposit in the bank; some of the former operators from the Chernobyl power station felt that some of those who worked for the Chernobyl Union had had nothing to do with Chernobyl and used the charity principally to arrange trips abroad for themselves and their friends.

However, the aim of the Chernobyl Union was not simply to solicit help from abroad or voluntary contributions from the Soviet people. At a meeting in the new town of Slavutich, built to replace Pripyat, Shovkovshytny described the law he and Yavorivsky had introduced in the Ukrainian Supreme Soviet, which would compensate all those who had suffered. The "safe" dose would be reduced from thirty-five rems to seven. Women who had lived in Pripyat would be permitted to retire on full pension at the age of forty-five, men at the age of fifty. Special passes would be issued to all the liquidators, who would be entitled to special rights: free travel on public transport, going to the front of any queues, and extra holidays up to eighty days. There would be financial compensation for all invalids among the liquidators, and the money would be paid from the Union budget. Moscow was responsible, so Moscow must pay; if it did not, the Ukraine would withhold the six or seven billion rubles that it contributed to the Union budget.

Mixed with these promises of concrete benefits were lofty thoughts on the human condition. "Remember," Shovkovshytny told the workers at the power station, "that our role on this planet is to create beauty and kindness so that the apocalypse will not happen. People must live and work in harmony together, and imitate God almighty. Christianity taught 'What is mine is yours.' Communism taught 'What is yours is mine.' If we can't keep all the commandments, we can at least keep one: love one another."

Then this preacher-politician took up his guitar and entertained his audience with some songs of his own composition.

6

The Chernobyl Law was passed, not just in the Ukraine but in Belorussia and the Russian Federation: however, its implementation had to await funds from the central government in Moscow. There had been no such delay in the appointment of a Chernobyl Commission, headed by Yavorivsky, to frame the Chernobyl Law, which employed a permanent staff in spacious offices in the middle of Kiev. Shovkovshytny served on the commission, as did Grodzinski. Later the Supreme Soviet established a Ministry for Chernobyl with Georgi Gotovchits as the minister. The Russians and Belorussians did likewise, so there were three ministers for Chernobyl, each with his staff of civil servants.

Ordinary people were not slow to appreciate that the radioactive cloud might have a silver lining. As soon as it became apparent that a liquidator's card would entitle its holder to certain benefits, there was a dramatic increase in the number of those who claimed to have worked in the thirty-kilometer zone. Those who were unable to claim the card envied the privileges given those who did. In Kiev there had already been considerable resentment that the evacuees from Pripyat and the thirty-kilometer zone had been allocated flats in the city ahead of those who had been waiting for many years. Now it was claimed that others had suffered from severe radioactive contamination—in the Urals, for example, after the accidents at Mayak. Nor was radiation the only form of pollution that had damaged the health of the population. Chemicals, pesticides and fertilizers had all seeped into the food chain, often causing more tangible harm than that claimed for radiation.

Even those who were in sympathy with the victims of Chernobyl asked whether the billions that were to be given in compensation would not be better spent on improving the health and nutrition of the population in general. The same applied to the vast cost of resettling up to a million people from areas where the levels of cesium and strontium would give them a lifetime dose of more than seven rems. In many cases the hazards of chemical pollution in the areas to which they were sent were greater than the dangers from radiation. What was the chemical equivalent of a rem?

Equally unquantifiable was the effect of the stress caused to those who were removed from their homes. Whereas actuaries in the West estimate that a person's life expectancy is reduced by a year each time he or she moves, Knijnikov estimated that the obligatory evacuation of a Ukrainian or Belorussian peasant from his ancestral village was the equivalent of a dose of more than one hundred rems.

However, the charges of genocide, once made, could not be withdrawn, and the fears, once they had arisen, had to be addressed. In Belorussia, Alexei and Antonia Dashuk, the sixty-year-old couple who had been left in their village forty kilometers north of Chernobyl, were finally moved from their large wooden house with its veranda, barn and cellar in 1989. Their dog and cat were left behind to run wild in the woods. They went first to Dubrovica, fifty kilometers further north, where they lived with Alexei's sister, who had earlier been evacuated from a village nearer to Chernobyl. There Antonia developed trouble with her liver and Alexei had a stroke and spent six weeks in the hospital.

They would have liked to stay in Dubrovica, where at least they could sit on a bench in front of the house and talk to passersby, but under the new, more stringent definitions enshrined in the Chernobyl Law, it too was considered unsafe for human habitation. They asked to be sent to another village, but the houses belonging to the collective were reserved for the working population. They might have bought a house, but it would have cost them twenty-five thousand rubles, and they had been given only fifteen thousand in compensation.

In 1990 the Dashuks were allocated a one-room flat in the suburbs of Minsk, with a certificate to establish their status:

> Alexei Mikhailovich Dashuk is from the settlement Rudiya of the Honicke region of the Gomel region, which is part of the zone of continuous control and zone of limited consumption of local foodstuffs as well as local plots of land, and this is to certify that they are evacuated according to the obligatory resettlement law to Minsk . . . in accordance with the decree No. 60b of the Central Committee of the Communist Party of Belorussia and the Council of Ministers on 21 March 1990.

The block was inhabited mostly by people who worked in the city. When it became known that the Dashuks had been evacuated from the contaminated zone, their neighbors regarded them with suspicion. When Antonia made friendly overtures to the woman next door, she shied away, saying apologetically, "I'm sorry, but I've got children." Despite their insistence that everything they had brought from their homes had been checked for contamination, the people who lived around them were afraid to enter their flat.

7

South of the thirty-kilometer zone, in the towns of Narodici and Polesskoe, the communities disintegrated as people waited to be moved from their homes. By the autumn of 1991, families with young children had already left, but in Polesskoe twelve thousand remained to be resettled. Eight hundred children remained in the school, which the year before had had over a thousand pupils. Asked at random, the children said they felt fine, but the local doctors reported that everyone's blood was affected by the radiation, leading to a decreased resistance to disease.

There were some changes for the better. The church had been re-opened and was now used for baptisms and marriages, even by the local party leaders, as it had been before the Revolution. There was also the extra allowance to buy clean food brought in from outside, but there was never enough; the shops were always empty, so people ate what they grew on their own plots of land. Radiation might be unhealthy, but so was starvation.

Worse than the material conditions, however, was the demoralization that had taken place among the inhabitants of Narodici and Polesskoe, who felt that they had been abandoned to their fate. In the autumn of 1991, a team of doctors from the Moscow Center for Intellectual and Human Technologies went to the towns and villages in the contaminated zone. Headed by Dr. Adolf Kharash and including Dr. Vladimir Lupandin, the group visited not just Narodici and Polesskoe but also Bryansk, Novozybkov and Vetka in Belorussia.

The conditions they found were appalling. There was some evidence of an increase of specific ailments that could be ascribed to radiation—seven children in Narodici with cataracts, for example—but far more serious was the moral degradation, which they ascribed to stress. People had become grasping and materialistic, indifferent to the fate of their neighbors. With only four thousand remaining from a population of sixty-five hundred inhabitants, there seemed no future whatsoever. The extra allowance paid for the purchase of clean food was spent entirely on drink—on vodka when they could get it, on wine when they could not. Wine cost 3.5 rubles a bottle, vodka 10 rubles, but thanks to the "coffin money," everyone could afford it. When the woman who ran the only liquor store received a consignment of vodka, she had to call on the militia to keep order.

Dr. Lupandin estimated that most of the inhabitants were now alcoholics. It was so widely believed that vodka washed out radionuclides that even the children drank. However, despite their anxiety about radiation, they continued to grow and eat their own food. Nor were the people deterred from having children. There had been no cases of congenital deformities, nor had any more malformed animals been born. The women continued to get pregnant and have babies, and when there were miscarriages, they were as likely to have been caused by alcoholism as by radiation. One of the best ways to help these people, Dr. Lupandin decided, was to build cafés serving soft drinks.

One of the principal reasons for the populace's demoralization was their sense of grievance against the local leaders, who they felt were exploiting the situation to their own advantage. There was a brisk black market in liquidators' cards and places at the top of the waiting list for new homes. Obliged by law to leave their old homes, ordinary families had been given insufficient compensation to buy new ones; only the party bosses received twice the real value of their homes. The medical supplies that came into the zone were resold to doctors from the big cities. Televisions and refrigerators, sent to Narodici to improve morale, were sequestered by the bosses and allocated to their friends. Everyone knew what was going on but could do nothing about it; anyone who protested would be charged by the militia with disorderly conduct.

All this might have been tolerable if the people had felt that they were going to have a better life, but the conditions where they were resettled were dire. The people of Narodici were allocated 950 flats in different towns, some as far away as Odessa, and 150 houses in a new settlement

called Brosilovka, sixty kilometers south of Narodici. For the unproductive old people, who were given the flats, it meant losing contact with
their families and their rural way of life. Like the Dashuks in Belorussia,
they had been used to a community where everyone was known to
everyone else, and where they could sit out on their verandas and watch
the world go by. The prospect of incarceration in a tower block was
often worse than remaining in an area of radioactive contamination.

For the working population, the move to Brosilovka seemed little
better. The land they were given to farm was swampy, infertile and five
kilometers from the village. The school and kindergarten were not much
closer. They complained that the new houses had been built in a hurry,
with substandard furnishings and crumbling concrete floors. The ancillary services were bad; the only shop was a small trailer that sold barley
coffee, pickles, tea and meat at ten rubles a jar. The bread came from
Zhitomir, fifty kilometers away, and milk was delivered every day, but
other items were in short supply. It was hard to get grain for their
chickens. There was no wood for their stoves; some brought it from
Narodici, even though it was contaminated; there was coal but it gave
off a filthy smoke. They had been promised gas to heat their homes, but
it had only been connected to the houses belonging to the party leaders.
The latter also had wooden floors to their houses and could get hold of
furniture, whereas some of the children of the state-farm workers had to
sleep on camp beds. There was no doctor because the house allocated
for one had been given by a party boss to his chauffeur.

Listening to this long lament from the former inhabitants of Narodici,
the team from the Moscow Center for Intellectual and Human Technologies was taken aback. Gentle men, who might have stepped out of
a story by Chekhov, they had carefully recorded the complaints of a
crowd of angry peasant women; the men were either absent or silent.
Retreating to the home (with wooden floor and color television) of the
former secretary of the District Party Committee, they had to match the
evident distress of these victims with the relatively sumptuous conditions
in which they lived. Even with concrete floors, the new houses were
lavish by Soviet standards. To a Moscow academic who could never
hope for anything better than a small flat, it seemed enviable to have a
detached house with two storys, five rooms, a bathroom and a garden.
To complain that the central heating was fueled by coal rather than the

gas they had been promised seemed to be looking a gift horse in the mouth.

The conclusion reached by Dr. Kharash was that the inhabitants' distress had little to do with their living conditions but came from the trauma of the resettlement. In Narodici their main anxiety had been the danger from radiation; now that this danger had been removed, they were left with the misery of having been uprooted. Although the women admitted that their children seemed healthier now and that their husbands had stopped drinking since going back to work, they still felt that no one cared about their suffering, and that they had been forgotten by the outside world.

Back at Narodici, a new wave of resettlement had started. On a bluff of land outside a beautiful wooden cottage with views of damp green meadows, slow-flowing rivulets and copses of birch trees, a woman and her crippled son waited to be picked up. Her husband had already left with the furniture for a village two hundred kilometers away in the Zhitomir region. She wept as she waited; her home was so beautiful that she would have liked to stay. No one liked the new village, and the compensation they had been offered was only a third of the cost of a new house. "Thirteen thousand rubles? What can one buy for thirteen thousand rubles? Even new furniture cannot be bought for that kind of money." If they had been paid proper compensation, she would have gone to live near her daughter; as it was, they had to go where they were sent and surrender any right to return. "They'll put us in that rat hole and leave us to rot." There were more tears. "Don't cry, Mama," said her crippled son. "Please don't cry."

Fifty kilometers away, well within the thirty-kilometer zone, Ivan and Irina Avramenko had never left their home. After hiding from the militia in the weeks following the accident, they had come out into the open and continued to live as before in the years that followed. Seven of their neighbors had done likewise, and within a couple of years another fifty people had returned to the village. In nearby Ilincy, there were as many as four hundred people. They were largely self-sufficient, living on their own cucumbers and potatoes, with mushrooms from the forests. Every now and then, officials came to try to persuade them to leave, but they countered by asking if there was anyplace in the whole Soviet Union that was unpolluted.

"If you really want to help us," said Irina to visitors from Kiev, "reconnect the electricity, organize deliveries of tobacco and bread, and of sugar to make vodka." When a people's deputy told them that the new Chernobyl Law would pay them compensation, they replied that in their experience people's deputies always cheated and told lies. "Soviet power is built on lies," shouted old Irina with a toothless cackle. "What have we known in our lives? Only war and famine. Isn't life just a vale of tears, as the Pope says? What do we need, anyway, but half a hectare of land while we're living and two square meters when we're dead? Do we worry about our children? Of course we worry about our children, for deserting the villages where their families have lived for hundreds of years. And if it's really so dangerous, why are party officials returning to their dachas? And building new roads to reach them? Why are they felling trees and taking them from the zone?" A further toothless cackle; then, Ivan in boots, Irina in slippers, they shuffled away.

"It's strange," a French journalist was heard to remark. "For the first time since I've come to the Soviet Union, I've met people who seem relaxed and happy."

XIII

1

By the autumn of 1989, it had become clear to the Soviet leaders that
on the question of Chernobyl they no longer enjoyed the trust of their
own people. In the elections that year, not only the radical environmentalists of Green World but also the Communist candidates ran on an
antinuclear ticket. The eminence and experience of the nation's leading
scientists counted for nothing. The reckoning now had to be paid for
decades of ideologically inspired distortions and outright lies.

The demands of the nationalists and environmentalists for the resettlement of everyone living in contaminated territory had grave implications for the Soviet economy. From the start it had been understood that
the accident would cost the state considerable sums of money. In
September 1986 Soviet Finance Minister Gostev estimated the cost at
two billion rubles. Included in his figure were the loss of electricity from
the fourth unit, the fourth unit itself (four hundred million rubles), the
partially built fifth and sixth units, other enterprises around the power
station, and the whole town of Pripyat. There were also the collectives
and state-owned farms whose land had been taken out of production,
the crops and livestock that had to be destroyed, compensation for
evacuees, new houses and the open-ended expense entailed in cleaning
the contaminated zones.

By March 1990 this estimate had been increased by a factor of two
hundred. Yuri Koryakin, the chief economist of the All-Union Research,
Design and Development Institute of Power Engineering, calculated
that by the end of the century the accident will have cost the Soviet state

between 170 and 215 billion roubles. The single largest item was the loss of produce from the fields and forests of the contaminated zones. Next came the loss of electricity production, not just from the fourth reactor at Chernobyl but also from the thirty-two other nuclear reactors that had been closed down or abandoned in the wake of the accident. Safety measures taken in the nuclear reactors that remained on line had cost from four to five billion rubles, leading to a 9 percent increase in their charges. Koryakin estimated the cost of cleaning the zone at between thirty-five and forty-five billion rubles.

Alongside the estimated lifetime limit of thirty-five rems that had been established by the government as the level for obligatory resettlement, a ground-contamination level had been set at forty curies per square kilometer. Now the proposals of the republican parliaments were that the first should be reduced to seven rems and the second to fifteen curies; indeed, some said that no one should be obliged to live where there was any contamination whatsoever. But to apply the lower levels would cost many million rubles. In a state that was virtually bankrupt, these demands had to be measured against other equally pressing, even urgent, calls upon its resources—for medicine, for example, and even for food.

Quite apart from the question of the cost of rehousing possibly a million more people from affected areas, the government knew that the stress caused by evacuation was more injurious than the effects of radiation. All attempts to convey this, however, were presented by RUKH and the Greens as an excuse to save money for the state, symptomatic of the Communists' callous indifference to the welfare of the people.

In desperation, the Soviet leaders asked for help from the international community. In October 1989 the government asked the International Atomic Energy Agency in Vienna to carry out "an international experts' assessment of the concept that the USSR has evolved to enable the population to live safely in areas affected by radioactive contamination following the Chernobyl accident, and an evaluation of the effectiveness of the steps taken in these areas to safeguard the health of the population."

It was not the first time that the International Atomic Energy Agency had been brought in to reinforce the credibility of the Soviet state. Only a fortnight after the accident, its Swedish director general, Hans Blix, and the head of its department of nuclear safety, Morris Rosen, had flown over the damaged reactor and subsequently reported their findings to a press conference in Moscow. Blix had said that he was "satisfied with the volume and nature of information" given him by his Soviet colleagues; Rosen had felt that the contaminated zone "will be inhabitable again."

Blix had also attended the conference on the medical aspects of the accident held in Kiev in May 1988, and in addition a team of experts from the World Health Organization had gone to the area in 1989. Their report had rebuked "scientists who are not well versed in radiation effects" who "attributed various biological and health effects to radiation exposure. . . . These changes cannot be attributed to radiation exposure, especially when the normal incidence is unknown, and are much more likely to be due to psychological factors and to stress."

However, the Western media, like the democrats in the republics, paid more attention to the claims made by scientists like Vladimir Chernousenko, a physicist from Kiev, who had been the scientific director of the task force set up by the Ukrainian Academy of Sciences to rectify the consequences of the accident. His claims that between seven thousand and ten thousand of the liquidators had already died as a result of the effects of radiation were given extensive coverage and were widely believed.

In response to the plea for help from the Soviet government, the International Atomic Energy Agency formed the International Chernobyl Project to assess the radiological situation in the affected areas. It called for help from different international organizations—the European Community, the United Nations Food and Agricultural Organization, the International Labor Office, the United Nations Scientific Committee on the Effects of Atomic Radiation, the World Health Organization and the World Meteorological Organization—and formed an International Advisory Committee of leading radiation experts from all over the world under the director of the Radiation Effects Research Foundation from Hiroshima, I. Shigematsu.

To discover the precise nature of the anxieties of those still living in the contaminated zone, a team of ten Western scientists went on a fact-finding mission to the republics between 25 and 30 March 1990.

At crowded meetings in town halls, they were asked questions that revealed the profound mistrust of the Soviet experts. "Is it safe to live here?" was the first question put by an inhabitant of Polesskoe. Then: "Why are we meeting in a small room when there is a larger one available in the town?" In Ovruc: "Is mother's milk being investigated at all, and could it potentially cause harm to an infant?" In the village of Novozybkov, in Russia, an old lady asked whether they should trust the Soviet or the Belorussian experts, who gave diametrically opposed advice. People wanted to know whether such disparate ailments as headaches, nosebleeds, weak legs, rotten teeth and gynecological infections could be ascribed to radiation. "Can we really live here any longer?"

2

Upon its return to Vienna, the International Advisory Committee drew up a work plan for a multidisciplinary team of international experts. Their brief was to examine critically the extensive information already available concerning the accident, the subsequent contamination by radiation, the effect this had on the inhabitants of the affected regions and the measures taken by the authorities to protect them. Excluded from the remit was the health of the 600,000 liquidators now scattered throughout the Soviet Union.

Time was of the essence: the team had to report back within a year. It was therefore impossible to check all the information available to them or embark upon a new and comprehensive assessment of all the consequences of the accident. In any case, there could be no independent evaluation of the radiological situation immediately after the accident because of the decay of the short-lived radioisotopes.

Between May 1990 and January 1991, 220 scientists, mostly from the United States and Western Europe, went to the affected regions of Russia, Belorussia and the Ukraine, and then returned to their own laboratories in the West to check the measurements they had taken. A year later, the report was ready. Acknowledging that it was difficult to be precise about "how much radiation dose has already been received or will be received," the project experts nevertheless made their own estimates which were two or three times *lower* than those made by Ilyn's Institute of Biophysics in Moscow. The contamination of food and drinking water was in most cases found to be significantly below the guideline levels established for international trade "and in many cases were below the limit of detection."

Comparing the health of the inhabitants of villages in the contaminated and nearby uncontaminated zones, the international team found that both were equally bad. Between 10 percent and 15 percent were in need of medical treatment of one kind or another, but none of their ailments could be ascribed directly to radiation. Children were found with low hemoglobin levels and low red-cell counts, but there was no "statistically significant differences between values for any age group of those examined in surveyed contaminated and surveyed control zones." The same was true when leukocytes and platelets were examined: there was no difference between children in the different zones, and "no significant variation from the data of other countries. . . . The immune systems of those examined . . . do not appear to have been significantly affected by the accident."

It was the same when it came to toxic elements such as lead and mercury. It had been feared that the lead dropped on the reactor might have been carried up into the atmosphere and subsequently inhaled by the inhabitants of the contaminated zones, but the levels of lead and mercury were much the same as those in the United States: the level of cadmium was considerably less; and all three were considerably less than those found in Italy, for example, or the Sudan.

There was no evidence of radiation-induced cataracts, and project results for thyroid sizes and size distribution and thyroid nodules in those examined were similar to those reported for populations in other countries. Nor did the dietary restrictions imposed in the wake of the accident appear to have affected the growth rate of children, which was well within published USSR and international norms. Adults, on the other hand, were generally overweight.

The report found high rates of infant mortality, but these had existed before the accident and had declined since. No statistically significant evidence was found of an increase in fetal anomalies as a result of radiation exposure. There might be in the future, as there might be an increase in cancer, but they would be difficult to detect even with a large and well-designed long-term epidemiological study. The only "statistically detectable increase" in the future might be the incidence of thyroid tumors and, possibly, leukemia.

3

Although it was humiliating for loyal Leninists like Ilyn and Guskova to have the deficiencies of Soviet health care exposed to the world, they were at least acquitted by the report of the charge of genocide. The charge of the Western experts was rather that they had done harm by *overestimating* the doses. "The cautious approach adopted . . . had two important negative consequences: firstly, the radiological consequences of continuing to live in contaminated areas were overrated and this contributed to additional and unnecessary fear and anxiety in the population; secondly, and more importantly, people will be relocated needlessly."

Moreover, the Soviet methodology had been wrong. Resettlement should not be based upon an estimated lifetime dose but rather on the additional dose likely to be averted. There was no precise figure above which radiation was dangerous and below which it was safe. To change the criteria from forty curies per square kilometer to fifteen reduced the likely lifetime dose by up to a mere eight rems. The consequent reduction in the risk of eventually contracting cancer was minute, yet the cost of resettlement was enormous and "seemed disproportionate to the effective radiological protection achieved by relocation." The "adverse health consequences of relocation should be considered," said the experts, "before any further relocation takes place."

When the findings of the report became known in the now disintegrating Soviet Union, they provoked outrage among the radical nationalists and environmentalists of RUKH and Green World. They were considered invalid because the project had not included the 600,000 liquidators. The experts had discounted the dangers of "hot particles" and removed "hot spots" from their calculations of average contamination. They had also chosen to examine the health of the affected population in the autumn when it was at its best, rather than at the end of the long months of winter.

More generally, the report was dismissed as a whitewash by the atomic-energy "mafia," who had a vested interest in allaying anxieties about nuclear power. To Lubov Kovalevskaya, they were afraid that Chernobyl would turn out to be the spark that lit a blaze against the entire industry; to her friend Dr. Lupandin, the report was a false, unprofessional piece of work. It was said that the international scientists had been bribed with free board and lodging at hotels in towns like Gomel and Minsk. In the debate in the Supreme Soviet on the Chernobyl Law in early May, Ilyn heard it said that to pay for the project, the Soviet government had expropriated one and a half million dollars from the Chernobyl charities.

Between 21 and 24 May 1991, more than five hundred scientists from all over the world gathered in Vienna to discuss the findings of the Chernobyl project. It came under fire from both the old Soviet scientific establishment and from the scientific champions of the new republics. Ilyn felt aggrieved that his intervention levels had been considered too cautious. "Let us look back and imagine the position of Soviet scientists who were responsible for the problems of standardization and the establishment of intervention levels at the time of the accident. If you put yourselves in this position, then you will understand that we didn't have any alternatives and that we were bound to be conservative in the philosophy we adopted for the introduction of the various parameters in the calculations which we made."

The principle which had guided them, he said, was the need to provide the maximum protection for the affected population, and to exclude all possible underestimates in their calculations. When it came to the dose to the thyroid, it had proved impossible to make accurate measurements. His colleagues in the Ukraine had done their best, but the teams of civil defense workers had proved incapable of making such measurements. "This was no fault of the scientists, but rather the fault of our system."

Professor Ramzayev, the persecuted scientist from the Leningrad Institute of Radiation Hygiene, was offended by the suggestion of an American, B. W. Wachholz, from the National Cancer Institute in Bethesda, that the Soviets had withheld data on the dose of iodine 131. "I can tell you," said Ramzayev, "that we were not asked for the information which you say was not presented. In particular, none of us was asked for data on iodine. They in fact exist. I want to clarify this so that the impression is not given to those present and to the world at large that the USSR did not present something that should have been presented. At least this was not the case in the Russian Federation."

Another American, L. R. Anspaugh from the Lawrence Livermore National Laboratory in California, tried to calm him. "We certainly did not mean to imply that any data were deliberately kept from us, but only that we did not have the basic measurements that went into the thyroid calculations in the sense of the very raw data on the thyroid measurements. . . . Our only comment was that, perhaps only because we did not ask for them directly enough at the time, we did not have the raw data on the actual exposure measurements of the thyroid itself that were made in the USSR."

Ilyn, however, pressed the international experts to estimate what proportion of the overall dose came from the short-lived radionuclides like iodine 131. "The question is important to us for the simple reason that a number of people in the USSR, including people who claim to be scientists, are attempting to undermine the reliability of our dose commitment evaluations by maintaining that we did not take these doses into account. . . . Therefore, I would like it to be stated more clearly, particularly since scientific representatives of our Republics are also here today." He was answered by a member of the United Nations Scientific Committee on the Effects of Radiation. "From available data, UNSCEAR has estimated the total transfer factor to be . . . approximately six percent of the total dose."

By and large, however, the Soviet scientists felt vindicated by the findings of the project. "We are happy," said Guskova, "to find out that the people we have been so worried about have not suffered any notable damage over these first five years, and that the doses are lower than we could have supposed." But one man's meat is another man's poison, and the project's findings were vehemently rejected by scientists from the republics of Belorussia and Ukraine.

At the very start of the conference, Professor Konoplia from Minsk produced new data to show that the atmospheric contamination was greater than the project scientists supposed. They had made their measurements at the wrong time of year. In spring and autumn, the months of agricultural activity, radioactive dust was transported outside the contaminated zones; and unforeseen events, like forest fires, increased the atmospheric content of radionuclides in the locality many times over.

Konoplia would not even discuss whether or not there had been an increase in illness in the affected zones: "Most people admit that there has been a rise." The vice president of the Ukrainian Academy of Sciences, V. G. Baryakhtar, told the conference of the various ailments that had affected the liquidators who had been left out of the project's terms of reference. He also insisted that "we have definite evidence of chromosome aberrations in children" and so "we should be a little more cautious about our conclusions." The director of the Kiev Institute of Endocrinology reported a rise in the number of thyroid cancers: twenty had been operated on in Kiev in 1990 as against one or two in the years 1985–89.

A Polish specialist, Professor Nauman, supported his colleagues from the republics that bordered his country: he thought the project's statement that "no health disorders could be attributed to radiation exposure" was unsound. It was impossible to be certain about the stochastic effects of radiation before 1995. He also clashed with the American expert, professor F. Mettler, a tall, gangling scientist from the Department of Radiology at the University of New Mexico, on the question of genetic abnormalities and "the conclusion you reached that there are no visible results of radiation in your study. I should like to ask, isn't it true that you have no real control group with which to compare your malformation rate? Isn't it true that genetic changes should be expected only ten, twenty or thirty years later, as you said, and isn't it true that your cytogenetic study showed that the background numbers are not sufficient for any conclusion to be made?"

Mettler agreed.

"If you did not find any changes related to the radiation," Nauman persisted, "you cannot in fact make any conclusions at this moment."

"Are we talking about malformations now?"

"I am talking about malformations, about genetic changes, and about the cytogenetic study. In fact, for malformations you had no control group from previous USSR studies from 1987 which would show any difference because, as you said, there are no data."

"For malformation rates in some areas there are previous data, and they are not statistically different."

"Yes, but you told us that intelligence quotient and the head size are most important and that you did not have data on them."

"That's right. I agree with you completely that intelligence quotient is interesting and that there are no background data."

"Do you agree that there is no reason to arrive at any conclusions concerning long-term genetic changes on the basis of the time elements in our study?"

"Absolutely."

"Thank you."

"I think you will know the answer to recessive genetic changes," concluded Mettler, "seven generations from now."

A. J. Gonzalez, an Argentinian from the International Atomic Energy Agency, was unhappy with the result of this exchange. "I feel that Dr. Nauman's intervention," he said, "could leave us in a state of uncertainty on the genetic impact of the accident of the accident, and this is something with which I personally do not agree. It is not true to say that we cannot draw any conclusions about hereditary effects. We know a lot about genetic effects already. We know that in all the epidemiological studies we have done (with only one exception, which we can discuss separately) there is no statistical evidence of hereditary effects. We know that the hereditary effects we are looking for have a very low probability of occurrence and a tremendously high background rate."

4

With growing dismay, the Western scientists who had worked so hard on the Chernobyl project saw a hemorrhage of trust in their findings. Not only were their reassuring conclusions about the likely genetic effects doubted, but the criteria they had established for relocation were questioned. "I have a moment ago received an unofficial translation," said Dr. Gonzalez, "of an article from *Izvestia* of the day before yesterday which I believe is very relevant to today's discussions. It states that . . . a new Soviet law was published that would establish the following limits for the relocation of people living in areas affected by radiation. It seems that (a) obligatory relocation with full compensation for loss of property should be considered when the average dose exceeds 5 millisie-verts (0.5 rem); (b) the population in areas with doses exceeding 1 millisievert (0.1 rem) could stay on or be relocated upon request, also with full compensation for property loss; and (c) people in areas with less than 1 millisieverts (0.1 rem) per year would have special status and be entitled to some bonuses and compensation without relocation.

"I ask myself what the consequences of this law would be if the USSR authorities decided, for instance, to control radon exposure in dwelling houses, following a similar policy. The average global dose for radon assessed by UNSCEAR is in the order of 1 millisievert (0.1 rem), and I can imagine that in the cold climate of northern USSR, with airtight houses, the levels can be higher than that."

Other contributors confirmed the absurdity of the new Soviet mea-sures. "There are many countries where a large proportion of the population receives much higher doses than 5 millisieverts (0.5 rem) per annum from radon and its daughters. Governments and populations affected by this radon irradiation do not bother about the situation." Yet under the new Chernobyl law, 5 millisieverts (0.5 rem) was now consid-ered the level for mandatory relocation, which, as Dr. Belayev from the Kurchatov Institute put it, amounted to "a violation of human rights."

There were also other far more lethal causes of pollution. "I would like to add something concerning cost and benefit," said Professor Jovanovich, a Yugoslav-American from the department of physics at the University of Manitoba. "We understand what cost is and we under-stand benefit as a benefit to the health of a population. Whenever I, as a nuclear physicist, come to a gathering like this, we always talk about the nuclear physics, about the radiation; we never talk about the air pollution. Yet two weeks ago, at an international conference in Anaheim, it was reported that air pollution in American cities has accounted for an estimated sixty thousand deaths a year, making it among the nation's top killers. This is not a crackpot report and it is nothing new. . . . Now if I have ten billion dollars or ten billion rubles and want to improve the health of the population, I have to decide whether to spend that money on relocating people because of the one hundred millisieverts or on cleaning up smokestacks. We should not forget that we live in a real world, and that there are other technologies more dangerous to the health of populations than nuclear power, including the Chernobyl accident."

Professor Hedemann Jensen, a Danish physicist, spelled out the view of the international experts in unambiguous terms. "Our conclusions are very firm. We say that there can be no justification on radiological protection grounds for the adoption of a more restrictive policy if consideration is limited to the cost and risk reduction alone." A bluff Scotsman, Dr. Ken Duncan, at one time chief medical officer of the British Radiological Protection Board, emphasized that it had been established beyond all doubt that any further expenditure on resettlement or compensation was a waste of resources. "People say, emotionally, and so much emotion comes into this, that you cannot put a monetary value on human life. In one sense you cannot and in another sense you must, because people also say you cannot spend too much on health. What absolute nonsense! You could spend the country's whole resources on health and you would be very much worse off at the end. Only the doctors would be very much better off."

What the scientists faced over Chernobyl, however, was not so much an epidemic of hypochondria as a widespread mistrust of science. "It is customary for doctors and scientists," said Duncan, "when they are faced with the unpleasant publicity which this field often attracts, to turn round and savage the nearest media person, whoever it may be. It is not entirely fair to blame journalists, broadcasters and people of that sort because they have been given a pseudoauthority by pseudoscientists in very many cases. . . . There is a great burden of responsibility on scientists to speak only in scientific terms. Not reputation, not money, but only honesty matters."

"Mankind has fallen into the trap that mankind itself has set," said Ilyn's colleague from the Institute of Biophysics, Dr. Buldakov. "Experts know that, among the various harmful factors, the radiation factor is perhaps the least harmful. But this is not known to the public at large, who do not want to reconcile themselves to the fact that for fifty years they have been fed falsehoods. They hold on to their firm belief that any radiation that is slightly higher than the background is fatal. It is very difficult to make them change their minds. . . . It is very difficult for us to prove . . . what nobody wants to believe."

. . .

Besides this philosophic irony, that there could be superstition within a science, there was also the historic irony that the party line in the Soviet Union, which in 1986 had been to play down the gravity of the accident, was now to exaggerate its consequences. Thus the Ukrainian minister for Chernobyl, Georgi Gotovchits, rejected the findings of the International Chernobyl Project. He said, "We feel that some of the basic conclusions that have been drawn concerning the radiological consequences of Chernobyl are too optimistic and could therefore be deleterious not only to the cleanup plan but also to nuclear safety problems in general."

This reversal in attitude was both baffling and frustrating for the Western experts who had worked so hard in an attempt to arrive at the truth about Chernobyl. For three years after the accident, the Soviets had refused to allow any scrutiny of the measures they had taken. Finally they had agreed, and the Western experts who had set about the task were prepared to discover ideologically inspired mendacity in the findings of scientific apparatchiks like Ilyn and Israel. "We were told," said an alert and eloquent physicist from Salzburg, Dr. Steinhäusler, not to "believe the figures you are given; don't trust the published map. . . . That was the situation in March 1990. Then one is confronted by people who are worried or frightened, who do not dare eat and drink what they produce, and this is happening in an area covering tens of thousands of square kilometers and involving hundreds of settlements. And behind you, there are seventy-one colleagues who are trying to find the truth, who are asking themselves, 'Is this map, given to us by our Soviet colleagues, correct? Are the milk and food data correct? Do they know how to measure? Do they know how to analyze? Is it safe for those people to live in that area?' The only way to answer these questions is to go out and measure, to go to the laboratory and find out how they do it, ask them to collaborate with you, give them unknown samples to measure, which is about the hardest test you can give a scientist. Our Soviet colleagues did this, and then put all the data together. The end result was [that] the maps are correct. They are not perfect; they could not be perfect. They were made under enormous time pressure and with constraints. I wish I could say that we could have done better, but I don't think we could.

"Our report is not a whitewash, with everything clean and perfect. Anyone who reads the technical report or reads the recommendations and conclusions does not have to read between the lines. It is stated quite clearly where there are areas that need improvement. . . . But the main conclusion does not change. It is: Yes, our Soviet colleagues know what they are doing.

"How does this help the people in the settlements who asked us, 'Can we eat the food? Can we drink the water?' I don't think it helps them if we state repeatedly, against the facts of measurement, that food is contaminated, soil is poisoned, water cannot be drunk, when the measurements clearly indicate the contrary. In most of the water measurements, for instance, we could not detect any radioactivity. Not because we used unsuitable equipment, but because there isn't any. . . . This has to be stated and spelled out very clearly. We can help these people not by reviving their fear but by believing in the measurements our Soviet colleagues made and that have been officially corroborated. This is the only way we can give these people back the trust that they need.

"There are indeed problems in the environment that we did not want to cover up, or make less of, such as the aquatic environment, which has a long-term potential [as] a problem, and we have indicated this. Radioactivity does not disappear overnight; it is in the sediment. We have stated this, we have measured it, and our Soviet colleagues know it. It can show up in fish and in other components of the aquatic environment. We did measure radioactivity in food, but we made it quite clear that commercially available food is under very good control. It is privately produced food, [that is,] food collected against official advice is dangerous. Soviet scientists cannot eliminate cesium from mushrooms grown in forests, and neither can anyone else. They can only tell people not to eat mushrooms.

"In summary, I would say that the environment does show radioactive contamination. Our technical report does not diminish the levels, but puts them into perspective and on an objective, numerical basis that the outside world can scrutinize and check. That is the only way science can progress, not by rumors and certainly not by frightening people in the affected areas. Finally, I think it is very unfair to a large segment of the scientific community—in this case, our Soviet colleagues and the other one hundred and ninety-nine members of the team—to suppose that we have been trying to cover up facts or whitewash the situation.

We have not. We have been trying very hard to give a true picture, and I think our Soviet colleagues have been trying to do the same."

5

Unfair it may have been, but to the enemies of the Communist party in the Soviet Union it was a small measure of injustice when compared to what they had suffered over the past fifty years. Even as the five hundred scientists were conferring in Vienna, the campaign was under way for the presidential elections in Russia, the first time in their history that the Russian people had ever been asked to choose a leader. The democrats' candidate was Boris Yeltsin, the Communists' Nikolai Ryzhkov.

As he always had been, Ryzhkov remained the most acceptable face of the old regime; nevertheless he epitomized that regime to all those who loathed the Communist system, and it was he who had led the Politburo commission in the aftermath of Chernobyl. In the impassioned atmosphere of the election—the extraordinary historic chance that few had thought they would ever live to see, to show through the ballot box what they thought of the system—it was asking too much of a long-oppressed people to agree with Dr. Steinhäusler that "our Soviet colleagues know what they are doing." Ryzhkov had led the government's efforts; Ryzhkov was a criminal and therefore must have been both morally and scientifically wrong.

Chernobyl was not an overt issue in the Russian presidential election, but just as the ecological movement had proved to be the wooden horse in which the anti-Communists had penetrated the walls of the Soviet Troy, so the charge of a cover-up on Chernobyl had become the clarion call with which a small band of democratic intellectuals had aroused the otherwise docile and conservative country people in Russia, Belorussia and the Ukraine.

On 12 June 1991, little more than a fortnight after the end of the conference in Vienna, the Russian people delivered their verdict on the Communist regime. Standing against five other candidates, Yeltsin won an absolute majority on the first ballot with 57 percent of the popular vote. The percentage of people who voted for Ryzhkov was only 16.9.

XIV

1

After his defeat in the Russian presidential elections, Nikolai Ryzhkov retired from political life. Yegor Ligachev had preceded him; after being overwhelmingly rejected as deputy secretary general at the Twenty-eighth Congress of the Communist Party in July 1990, he had announced his retirement to his native Siberia.

But their departure did not mean that the conservative forces in the party and government had been routed. Mikhail Gorbachev, with a new team of conservative ministers, could still call upon the vast coercive forces of the ministries of defense, the interior and the KGB. Few imagined that the heirs to Bolshevism would give up their inheritance without a fight.

The powers of patronage remained with the Central Committee, so those who did not earn their living either as people's deputies or as radical journalists had to beware of biting the hand that fed them. Nonetheless, the election of Boris Yeltsin revealed a dramatic change in people's attitudes toward the Socialist system: and this same liberation from the ideological outlook that had been instilled in them in nursery school, secondary school and college, in the Pioneers, the Komsomol and the party, had affected many of the protagonists in the tragedy of Chernobyl.

Lubov Kovalevskaya, whose article in *Literaturnaya Ukraina* about the shortcomings in the construction of the nuclear power station had been published shortly before the accident, had been acclaimed as a prophet by the media in the West. Reporting on conditions in the zone in the years following the accident, she had become profoundly disillusioned about nuclear power as such. She wrote, "The consequences of the Chernobyl tragedy are irreversible and eternal. Mankind will have to adapt itself to the post-Chernobyl condition. No rescue operations can prevent genetic contamination of the environment, change in blood formulae, genetic code, landscape, traditions and culture, degradation and insanity of nature and man as a part of it. The world is one indivisible whole."

Lubov also expressed commonly held doubts about the impartiality of the International Atomic Energy Agency: "The aim of the organization is to *develop* the nuclear power industry, so it is interested in covering up not only the problem of plutonium control but also acute problems of the nuclear power industry."

In 1989, when all information about nuclear matters was finally declassified, Lubov was able to describe the way in which the state's security apparatus had controlled the dissemination of information about the disaster, and she exposed corruption in the charities set up to mitigate its effects. "It is well known," she told Radio Free Europe in 1990, "that Glavatom [the State Committee for the Use of Atomic Energy] took sixty-five million rubles from account No. 904 and transferred it to its own account."

Unlike many of the radical democrats and Greens, however, Lubov Kovalevskaya did not drop one party line merely to adopt another. With no need to court their votes, she had the courage to tell former nuclear power station workers in the Chernobyl Union that they were partly to blame for the accident because they had failed to speak out against the corruption and inefficiency in the management of the station. She had an ambiguous relationship with Shcherbak and Green World. He had given her a creditable role in his book, but she saw his Green World movement develop a self-seeking intolerance of its own. In her view, it exploited the Chernobyl tragedy to promote its own nationalist agenda.

Sometimes those who answered the telephone at Green World refused to speak Russian—unless the caller was interpreting for a Western correspondent offering hard currency for an interview or some sensational statistics. As a Russian living in Kiev, Lubov found it increasingly difficult to get her work published; it was often returned with the recommendation that it be translated into Ukrainian. She was obliged to work for *Raguda,* the only Russian-language magazine left in Kiev, which was run by hard-line Communists who were inevitably unsympathetic to her point of view.

Like many others in the wake of Chernobyl, Lubov's disillusion with the Communist system led her to doubt what she had been taught about religion. She had been baptized as a child, but to confirm her return to religious belief she was baptized again in the Cathedral of St. Volodomyr in Kiev. Valentina Brukhanov went to the same church to pray quietly for her imprisoned husband. Both actions were symptoms of a widespread return to religious belief in the Ukraine. On the first anniversary of Chernobyl, in the village of Hriushiw in the Ukraine, a twelve-year-old child had a vision of the Virgin Mary. The accident was seen as an act of God, and even as a punishment for the atheism of the Bolsheviks. In Ukrainian, "chernobyl" was the word for a species of the bitter plant wormwood, mentioned in Chapter 8, verses 10–11, of the Book of Revelation: "The third angel blew his trumpet and a huge star fell from the sky, burning like a ball of fire, and it fell on a third of all rivers and springs; this was the star called Wormwood, and a third of all water turned to bitter wormwood, so that many people died from drinking it."

In Russia, as well, many returned to the faith of Orthodox Christianity. At Easter, the banners that had once proclaimed the triumph of the proletarian revolution now read CHRIST IS RISEN, HE IS RISEN INDEED. A ceremony held in the auditorium of the Olympic Village in Moscow to mark the fifth anniversary of the Chernobyl disaster was presided over not by a party leader but by the Patriarch of Moscow. "God help those who died at Chernobyl," he prayed, "and save the world from another tragedy of this kind."

For many raised as atheists, it was not easy to exchange their Marxist convictions for Christian beliefs. Moreover, there were Soviet citizens like the Tatar Davletbayevs whose traditional religion was not Christianity but Islam. Raised as Communists, and with no reason until Chernobyl to doubt its atheist philosophy, Inze had rediscovered through suffering her people's one true God. No longer did she pray to the mummified body of Lenin or to the bronze statue of Pushkin but took her son Marat across Moscow to pray to Allah in a Mosque.

In Slavutich, the new town built for the nuclear power station's personnel forty kilometers from the Chernobyl plant, agitation started for the construction of a church. The Communist mayor suggested a "ritual hall" in a museum to be devoted to the power station, but the new believers would not be fobbed off. Plans were drawn up for an Orthodox church, but by the time they had been approved there was no money to build it.

The morale of those living in Slavutich was low. Started on virgin territory in 1986, the town was going to be as fine as Pripyat. To show their solidarity with this All-Union enterprise, each republic built a sector; therefore there was a variety of architectural styles, from the Nordic houses built by the Baltic states to the Levantine blocks of flats built by Kazakhstan and Azerbaijan. There were the same wide walkways, parks and children's playgrounds and, as at Pripyat, the beautiful forests of the Polessia, which began at the boundaries of the town.

When the maps showing radioactive contamination were finally published in 1989, it became clear that the town had been built on contaminated land—between one and five curies per square kilometer. It was unsafe to walk in the forests and hazardous to eat the mushrooms. As a result, some looked for jobs elsewhere. Many of the former residents of Pripyat had been given flats in Kiev and had initially commuted from there to the power station. Great pressure had to be put on them and their wives to get them to live in Slavutich.

Despite the preference for Kiev of his cheerful and sociable wife, Ylena, the dutiful Vadim Grishenka stayed on as deputy chief engineer responsible for the sarcophagus. He had received a dose of thirty rems at the time of the accident; afterward he had ceased to count. Vladimir Chugunov had returned despite a dose of one hundred rems. Nikolai Steinberg remained as chief engineer until 1987, and his loyal secretary, Katya Litovsky, became secretary to the director. But the blight cast by the accident over the whole enterprise could not easily be dispelled. There were too many strange faces, and the memory of old friends—both those who had died and those who were in prison.

At the time of the trial of Brukhanov, Dyatlov and the others from the Chernobyl plant, there been an assurance that an investigation into the culpability of the designers of the RBMK-1000 reactors was continuing. However, no charges had been brought, and no further investigation was undertaken. In March 1987 Nikolai Steinberg left Chernobyl to become vice chairman of the All-Union Committee for Nuclear Safety in Moscow. Raised by Communist parents, he had been an active member of the Pioneers, the Komsomol and the party, but the disaster at Chernobyl had revealed to him the flaws in the ideology he had so uncritically accepted throughout his youth.

In 1989 Steinberg resigned from the Communist party. In May 1991, at the First International Sakharov Congress on Human Rights in Moscow, he presented a report on the true causes of the accident, which had the full authority of his official position. The root causes of the accident, he concluded, combined "scientific, technological, socioeconomic and human factors." In essence, in order to complete the tests on the turbines, the operators had brought the reactor into an unstable condition. The most serious error—made by Toptunov, on Dyatlov's instructions and with Akimov's acquiescence—was to raise the power of the reactor after it had dropped to zero, but this had only been done because the operators felt under pressure to complete the test.

In his report Steinberg wrote: "A refusal to go on with the tests meant a failure to fulfill the plan and a postponement of the verification of the important safety regime for many months. Under those circumstances, the unit operators and managers made a decision that, in all probability, predetermined the subsequent accident. Why 'in all probability'? Because there is as yet no answer to the question of at what moment the reactor could have been shut down without running the risk of a serious accident."

The personnel certainly violated the operating instructions, but they did not know and could not have known that the control rods of the RBMK reactor had a design fault whereby their insertion, instead of leading to an immediate decline in reactivity, led to a momentary increase—sufficient, given the condition of the reactor on 26 April, to lead to a runaway surge in power and subsequent explosion.

Only very complex calculations would make it possible to discover what actually happened that night, and "unfortunately . . . no representative model has been developed up to now." However, all those investigating the causes agreed that the power surge was initiated by the insertion of the control rods into the core. "Thus the Chernobyl accident comes within the standard pattern of most severe accidents in the world. It begins with an accumulation of small breaches of the regulations; more than ten were identified in the design of the safety system of the No. 4 reactor. These produce a set of undesirable properties and occurrences that, when taken separately, do not seem to be particularly dangerous, but finally an initiating event occurs that, in this particular case, was the subjective actions of the personnel that allowed the potentially destructive and dangerous qualities of the reactor to be released."

Given this context, Steinberg continued, it seemed unproductive to ask who was to blame, "those who hang a rifle on the wall, aware that it is loaded, or those who inadvertently pull the trigger." Rather, one should look to the more general causes of the Chernobyl disaster: the lack of specific legal accountability for the safety of nuclear power stations; inadequate control over the quality of both the construction and the operation of the reactors because the regulatory bodies (such as his own) had insufficient powers. "Even now, a substantial part of the nuclear-fuel-cycle facilities has not been placed under the supervision of the state regulatory authority."

Then there was the human factor: a lack of consideration for the psychological impulses that lead to human behavior (more specifically, the idea of an operator as an automaton), and at the same time a tradition of driving a worker on to such superhuman accomplishments as breaking production records or exceeding the norms, which encourage him to improvise and cut corners. "This list could be extended. However, most of [these factors] could be cumulatively defined as the complete absence of a safety culture." In other words, the accident at Chernobyl could be ascribed to the old Communist adage that the end justifies the means.

Complementing Steinberg's paper for the Sakharov Congress, Dr. Armen Abagyan also produced a report reaching similar conclusions on the technical questions, views he had always held but which hitherto it had been impolitic to publish. He also described the measures that had been taken to prevent such an accident from happening again: the introduction of additional control rods; reducing the time it took for them to descend into the core; the use of uranium with 2.4 percent enrichment; updating the documentation and the training of the personnel. Like Steinberg's report, Dr. Abagyan's findings did not exonerate the operators, but they at least demonstrated that the celebrated breaches of the regulations were not in themselves what had caused the accident, and they implicitly admitted that the operators could not have learned either from their training or from the documentation that existed at the time that the reactor could explode.

2

Anatoli Dyatlov was not satisfied. Released early from prison because of his deteriorating health, he embarked upon a campaign to prove his complete innocence. It was one thing to have been condemned by a Soviet court—he knew that the system required scapegoats—but it was quite another that the Western scientists who had compiled the INSAG-1 report in 1986 should have followed the party line. In a letter to its authors written toward the end of 1991, he claimed that they had been deceived by Legasov, and he refuted point by point their claim that mistakes made by the operators had led to the accident. He insisted that the regulations they were said to have breached had only been drawn up *after* the accident, and that the changes made to the RBMK-1000 reactor proved that its design had been inherently dangerous. The cause—not the main cause, not the root cause, but the *only* cause of the accident—was to be found in the characteristics of the RBMK-1000 reactor.

Those serving shorter sentences had been released before Dyatlov, and Fomin had been freed soon after because of his fragile state of mind. By the fifth anniversary of the disaster, only Brukhanov remained in jail. To some this seemed just, not because he had been responsible for the accident but because in the aftermath he had put the interests of the party before the lives of his own personnel.

However, Brukhanov was not without defenders in high places. Professor Anatoli Nazarov, one of only three doctors of ecological science in the Soviet Union, had been elected to the Congress of People's Deputies by the Academy of Sciences. A small man with a balding head and a beard, he had never been a member of the Communist party because he did not share its ideals. Even after his election to the Congress, he saw himself not as a politician but as a specialist of a kind much needed in the polluted Soviet Union.

Nazarov's first assault on the sacred cows of the military-industrial complex had been a pamphlet criticizing plans to build a new reactor in the Chelyabinsk region of the Urals. In it he described the nuclear pollution caused by the disastrous explosion at Mayak. This had been revealed in the West as early as the 1970s by the exiled nuclear scientist Zhores Medvedev, who, while still in the Soviet Union, by reading articles about the effects of radiation in Russian scientific magazines, had inferred that a serious accident must have occurred. Medvedev's conclusions had been dismissed as anti-Communist propaganda, even by nuclear scientists in the West, and had been kept as state secrets in the Soviet Union itself. Now Nazarov's pamphlet confirmed everything that Medvedev had supposed.

Despite threats and bribes, Nazarov continued his opposition to nuclear pollution and was therefore the natural choice to head a Supreme Soviet commission on Chernobyl. With the full power of the legislature at its disposal, it included among its investigations "an analysis of materials of Criminal Case No. 19–73," which suggested that it was the last of the Soviet show trials. This read, in part:

> There is no doubt that a number of serious breaches in the running of the reactor allowed by those convicted in this case can be considered as proved during the trial. However, what is the nature of such breaches? For example, a group of prominent specialists from Obninsk, led by Professor B. G. Dubovski, which carefully investigated the problem, came to conclusion that "if the safety protection of the RBMK-1000 reactor had been close to what it was supposed to be, then all the mistakes made by the operators would only have led to a week's shutdown of Unit 4 of the Chernobyl nuclear power station." Thus, the worst charge would have been *negligence*—i.e. a crime in line with Article 167 of the Ukrainian SSR Criminal Code.

The investigators went on to largely exonerate Brukhanov:

> It is difficult to consider as valid the conviction of the ex-director of the Chernobyl nuclear power station, V. P. Brukhanov, accord-

ing to Article 165 of the Ukrainian Criminal Code ("abuse of power or administrative position"). He was found guilty of intentionally hiding the fact that radiation levels were considerably above the norm; that he sent the operators on different missions into dangerous areas without providing them with necessary protective clothing, etc. First, there is evidence in the case testifying to the fact that very quickly numerous bosses arrived on the scene of the accident and that they appealed to him "not to create a panic." Second, it is doubtful that in the atmosphere that prevailed shortly after the accident, the director of the power station could have had any precise information. Academician Legasov stressed this particular point in his memoirs. Thus the order to go on missions to dangerous zones without adequate protective clothes to combat the consequences of the accident can be regarded as *dire necessity* under the circumstances that then prevailed.

Innocent or guilty, Brukhanov and his fellow convicts had suffered from more than their incarceration. Dyatlov had received a dose at Chernobyl of 550 rems; with a poor diet and rudimentary medical care in prison, he looked so old when he came out that it seemed he had aged fifteen years in only three. Brukhanov fared slightly better; he had a lower dose, and in his own estimation was better able to avoid the ill effects of radiation because he had spent his childhood in Tashkent under the hot sun.

But even with the best medical care, many of those who had received high doses were still suffering five years after the accident. Piotr Palamarchuk, who had defied all medical prognostication by surviving a dose of between 750 and 800 rems, had grown stronger as the years passed, but there were still open sores on his back and legs. Sasha Yuvchenko also had open sores on his arms and was flown to Bavaria for skin grafts by German army surgeons, which Dr. Baranov felt could have been done in Hospital No. 6 had everyone not lost faith in Soviet medicine.

It was a demoralizing period for the doctors at Hospital No. 6. It had been the boast of the Soviets that whatever the other shortcomings of their system, it took care of the health of its people, but the international experts had described the low level of health care for peasants in Belorussia and the Ukraine. At the time of Chernobyl, Guskova, Baranov and the young men and women working in Hospital No. 6 had, at great risk to their own health, done everything in their power for the patients in their care, only to be vilified by Professor Vorobyov as the lackeys of the Third Division and the KGB. To rub salt in these wounds, Baranov and Guskova were given a showing of an American movie, *The Final Warning,* based on the book by Dr. Robert Gale, in which the Russian doctors played minor roles while the American specialist was the hero.

Razim Davletbayev remained a semi-invalid: his hair was now gray, his movements slow. He had a nominal job at the Ministry of Atomic Energy but suffered from regular bouts of weakness and fatigue; even a common cold turned into a prolonged disease. Inze Davletbayev remained a close friend of Akimov's wife, Luba, who, while she seemed to Inze to be detached about their marriage while he was living, became the champion of her husband after he was dead.

Alexander Nemirovsky, one of the firemen who had received a dose of up to two hundred rems while laying the hose under the reactor, was rewarded with promotion and the sum of 700 rubles. He also developed a malignant tumor on his spleen. Ivan Shavrey, who was told after analysis in Israel that he had received a dose of six hundred rems, went back to work as a fireman because an invalid's pension was too small. Many of the wives suffered as well; Valentina Brukhanov was in the same hospital as Ylena Grishenka, who had multiple growths in her ovaries, while Lubov Kovalevskaya had nodules in her thyroid.

Worse than their own suffering was that which these people observed or anticipated in their children. The Palamarchuks' younger daughter, who had been in Pripyat at the time of the accident, had developed continuous headaches, though the doctors assured them that this had nothing to do with radiation. If the children were healthy, their parents waited and watched for the appearance of symptoms; if, like the Yuvchenkos' son, they were a little listless, they tried hard to remember whether or not the children had been like that before the accident. Certainly life in a small flat in a suburb of Moscow in times of shortages and political uncertainty compared badly to the healthy life they had all enjoyed in Pripyat. Piotr Palamarchuk felt much healthier when he visited his native Vinnitsa in the south of the Ukraine, but he dared not move away from the specialists in Hospital No. 6.

There were also fears for future generations. The risk of bearing a malformed baby made couples hesitate to have a second child. Still, Alla Kirschenbaum, the wife of the young turbine engineer from the fourth unit, gave birth to a second child five years after the disaster, and it was a healthy baby girl. By then she and Igor had decided that when the opportunity arose they would emigrate to Israel.

The men who appeared least affected by high doses of radiation were the military leaders—first and foremost, Major General Pikalov, commander of the chemical troops, now a Hero of the Soviet Union. Of all the 260,000 soldiers involved in the aftermath of the accident, Pikalov believed that he had had the largest dose. Guskova estimated an internal dose of eighty-seven rems and an external one of fifty-three rems. Radioactive particles had left ulcers on his legs. However, this had not stopped him from remaining in command of the chemical troops until his retirement, and after his retirement from working for a Ph.D. He also served as president of the Liquidators Association, and took an interest in the new political developments through his daughter, who was a democratic member of the local council. There were times when he felt weak or slept badly at night, but he had had frequent medical checkups and his blood formula was normal.

General Berdov, the commander of the Ukrainian militia received a dose estimated at one hundred rems with no notable ill effects. General Ivanov, the robust deputy commander of the civil defense, who had received a dose of about sixty rems, saw no reason to revise his view that a small dose of radiation did no one any harm. Indeed, for fruit and vegetables it could be positively beneficial, so why not for human beings too? The dose to his thyroid had been about 120 rads; as a result his metabolism seemed to have changed for the better. He had absorbed a whole cocktail of ruthenium, niobium, zirconium, and cesium. Some elements, like ruthenium, had a short half-life, but the cesium was more problematic. A couple of months after the accident, the director of a nuclear station told Ivanov that he could wash cesium out of his body with beer. The barley in the beer, the scientist insisted, absorbed big particles of cesium and came out with sweat. It was a therapy that appealed to the general, as did another theory, that red wine was good for the bone marrow and the spleen.

Others were less resilient than these military men. Victor Knijnikov suffered a heart attack toward the end of 1990. He recovered and ascribed it to stress. Boris Scherbina died of a heart attack in 1989, but like most others in the Soviet Union, he had been a chain smoker. Victor Koreshkov, the chief engineer of the Kiev metroconstruction workers, also died, as did seven thousand of the liquidators. To some, like the Ukrainian scientist Vladimir Chernousenko, this figure was proof of the hazards of radiation; to others it was the normal death rate for six hundred thousand people living in the Soviet Union at the time. The death of almost anyone who had been near the fourth reactor at Chernobyl came to be ascribed to radiation—for example, the Ukrainian film director Vladimir Shevchenko, who was ill before he took footage of the burning reactor and died a year later.

General Ivanov was particularly incensed at the publicity about the medical treatment in Seattle of the helicopter pilot Anatoli Grishenko. Grishenko had been one of the test pilots, but no one remembered his spending unusually long periods over the reactor. Certainly, General Ivanov insisted, he had never been made a Hero of the Soviet Union, as was announced in the U.S. press.

3

In 1989 a team of scientists from the Kurchatov Institute formed the Complex Expedition to investigate conditions within the sarcophagus itself. In the aftermath of the accident, no one knew what had happened to the nuclear fuel. Ever since the graphite fire had burned itself out on 6 May 1986, the temperature within the ruins of the reactor had gradually subsided. In theory, however, there was always the possibility that if a critical mass remained buried within the sarcophagus, fission might resume.

In 1989 the search for the fuel took the members of the Complex Expedition down into the basement of the fourth block. Working in areas of high radiation with inadequate protective clothing, they came across fractured fuel rods that were empty and cold. However, their instruments suggested a highly radioactive mass ahead of them. They sent in a television camera fixed to a robot to examine it and discovered what they called "the elephant's foot."

This was a once molten mass that had solidified as it had oozed out of the reactor into the foundations of the fourth block. Like a limestone deposit in an underground cavern, it had rippled layers of a substance that the team had never encountered before. The radiation level on its surface was ten thousand roentgens an hour. To discover what it was, the scientists called in a police marksman, who shot off fragments with an AK-47 rifle. These showed that the elephant's foot was a once-molten flow of nuclear fuel and sand.

Where had this lava come from? To find out, the Complex Expedition drilled a hole into the base of the reactor. Inserting a camera through it, they saw that it was now empty. At the base there remained some fragments of graphite, but there was no nuclear fuel. Returning to the basement beneath the reactor, they explored chambers adjacent to the emptied bubbler pool and suddenly came upon the source of the lava flow: a now-solidified mass of nuclear fuel and sand, bright yellow and black in the light of their lamps, and covered with strange new crystalline forms.

To Konstantin Checherov, the leader of the expedition, it was a wonderful sight, not just because it explained what had happened to the fuel but also because it had its own strange beauty. It also solved the riddle that had so puzzled Legasov in the immediate aftermath of the accident. It was known that the force of the explosion had blown the biological protection shield off the top of the reactor; now it was clear that it had also forced down the base of the reactor. In doing so, the sand lining the reactor had started to pour into the basement, mixing with the molten fuel from the core. In contact with such high temperatures, the sand had vitrified and then solidified, at the same time diluting the fuel and trapping it like a fly in amber. There would be no new chain reaction.

Knowing the whereabouts and the condition of the nuclear fuel did not remove all potential future hazards from the fourth reactor. Most acute was the condition of the sarcophagus. Built with urgency under atrocious conditions, it was supported by the structure of the fourth unit itself. This structure had been damaged by the force of the explosion and had shifted a little since the sarcophagus had been built. As a result there were gaps where it was open to the elements, where rain could enter and, washing through the highly contaminated contents, seep into the soil and contaminate the groundwater.

There was also a possibility that at some stage the sarcophagus could collapse onto the many tons of highly radioactive dust, sending a new cloud of radionuclides into the air and endangering the personnel at the station. The same thing would happen if the huge biological protection shield, now supported only by broken pipes, should fall to the ground. There was no obvious solution to this problem. It might be possible to rebuild the sarcophagus with stronger foundations, or to build a mammoth new one to encase the one already in place, but either solution would cost large sums of money, which the bankrupt country could ill afford.

Nor was it clear what would happen to the other three units. A resolution was passed in the Ukrainian Supreme Soviet to decommission the entire Chernobyl nuclear power station. However, the plant still belonged to the Ministry of Energy and Electrification in Moscow, which insisted that the economy could not afford to lose the three thousand megawatts of electrical power. The future of Chernobyl became one of many bones of contention between government in Moscow and those in the republics, and its fate remained, as it had been since 26 April 1986, inextricably entwined with the political evolution of the country.

4

On 21 August 1991, Mikhail Gorbachev was scheduled to sign a new treaty with the constituent republics of the Soviet Union. After Yeltsin's election as president of Russia, it was clear that some devolution of power must take place. According to the draft of the treaty, most internal matters were to be left to the republics. Most notably, they were to have control over their own mineral resources and could legislate to allow "all forms of property," even the private property abolished by the Bolshevik Revolution. As Soviet president, Gorbachev would be responsible for foreign affairs and defense.

Two days before the treaty was due to be signed in Moscow, and while Gorbachev was still on vacation in his state dacha in the Crimea, tanks and armored personnel carriers moved into Moscow. It was announced on Soviet television that Mikhail Gorbachev had fallen ill and that his powers had been assumed by a state committee headed by the vice president, Gennadi Yanayev. Included among its members were the prime minister; the minister of defense, Dmitri Yazov; the minister of the interior; the head of the KGB and also, most significantly, Oleg Baklanov, the chief of the military-industrial complex, so long concealed behind the euphemistic title of the Ministry of Medium Machine Building.

It was a preemptive strike by the heirs of Stalin and Beria against a treaty that would destroy the Soviet Union. Looking back on what Legasov had described as the "heroic" decades of their history, when the whole world had trembled before the might of the first Socialist state, these men hoped that their nostalgia would be shared by the populace. It was the coup that many had prophesied and more had feared. Theoretically the state committee had at its beck and call the largest coercive force in the history of mankind, but the patriotic programs that ran on television seemed like a historical pastiche harking back to the era of stagnation and beyond. The junta had nothing to offer the Soviet people but nostalgia for an era that for a majority had been a passage through hell.

The coup failed. First, Gorbachev would not cooperate. When the commander of his presidential guard at his villa in the Crimea announced unexpected visitors, Gorbachev's first thought was that there had been another Chernobyl. Discovering that it was merely a coup, he refused to sign a decree proclaiming a state of emergency and told the conspirators to go to hell.

Nor would Yeltsin submit. The democratically elected president of Russia remained defiant in his republican legislature, the "White House" overlooking the Moscow River, and thousands of Muscovites came to protect him with a human shield. Among them were the grown-up children of Nikolai Steinberg and bringing them food was the daughter of General Pikalov. The decent, honest Sasha Yuvchenko, watching these events on television at home and seeing the democratic mask drop from the face of the Communist party to reveal a self-interested oligarchy of self-deceiving old men, suddenly realized the extent to which his whole outlook had been formed by false indoctrination. On the third day of the coup he tore up his party card.

In the new town of Slavutich, the equally honest and courageous Vadim Grishenka, the taciturn man of few words married to Ylena, the woman of many, sternly refused to succumb to fashion; he still believed in socialist ideals and remained in the party. Throughout the Soviet Union, the same emotional turmoil went on in the minds of millions of men and women, but the success of the coup depended upon very few. Its leaders had no stomach for the kind of slaughter that had come so easily to their mentors, Stalin and Beria. Nor was the juggernaut of coercion what it once was. The young conscript commanders of some of the tanks sent to surround the White House were persuaded by the crowds to turn to defend it, and when ordered to storm the White House at midnight on 20 August, the crack assault force of the KGB, known as the Alpha Group, came to a unanimous decision to disobey.

By the morning of the 21st, its leaders realized that the coup had failed, and that afternoon the tanks were withdrawn from Moscow. Yazov and Yanayev were arrested. Boris Pug, Ryzhkov's successor as prime minister, shot himself. So, too, did Marshal Akhromeev, chief of the general staff at the time of Chernobyl, who had sided with the plotters.

A race now began to reach Gorbachev. The leaders of the coup hoped to persuade him to give some kind of legitimacy to what they had done, but Yeltsin's delegation, consisting of his vice president, Rutskoi, and his prime minister, Ivan Silayev (the same Silayev who had succeeded Scherbina as chief of the Chernobyl Commission), reached the Soviet president first.

Gorbachev returned to the Kremlin, but not to power. To the victor goes the spoils, and the victor was Yeltsin. While his supporters pulled down monuments to Lenin and the statue of the KGB's founder, Feliks Dzerzhinski, under the eyes of the KGB, Yeltsin declared the Communist party to be "the organizing and inspiring force" behind the coup. Invited to address the Russian parliament two days after the coup's collapse, Gorbachev was heckled and booed by Yeltsin's supporters, and asked why he had made ministers of the coup leaders. Yeltsin then invited him to read out the names of the conspirators.

"My situation right now is hard enough," said Gorbachev. "Please don't make it more difficult for me."

"Comrades," said Yeltsin "How about the decree to suspend all the activities of the Russian Communist party?"

"To prohibit the Communist party," Gorbachev protested, "I have to tell you, would be a mistake for such a democratic Supreme Soviet, for such a democratic president of Russia."

Yeltsin ignored him. "The decree is hereby signed," he said. "It's not a prohibition. It is a decree of cessation of activities . . . so this can be dealt with by the courts."

5

Three weeks later, on 10 September, Victor Brukhanov appeared before a judge in Uman, where he had been imprisoned, to consider his petition for early release. He was the last of those convicted for the accident in Chernobyl to remain in prison. Laushkin had died of stomach cancer soon after his release. Fomin was now working at the Kalinin nuclear power station, where his mental condition was still said to be strange. Rogozhkin and Kovalenko had gone back to work at Chernobyl, and Dyatlov was now receiving treatment in Germany.

Brukhanov had suffered much over the past five years—physically from the privations of a KGB prison; psychologically from the humiliation of his fall as the decorated director of the country's largest power station to a menial laborer among convicted currency speculators and black marketeers. Morally he could not escape the knowledge that while his conscience excused him from blame for what had happened, he was nonetheless nominally responsible for the suffering of millions of people, among them his devoted wife Valentina, ill and unhappy in a hospital in Kiev.

Brukhanov's time in prison had not been entirely wasted; he had acquired a good knowledge of English, and his spirit had not been broken. When he was asked by the judge whether he now admitted his guilt, he replied that he did not. "I have been punished for the mistakes of my subordinates," he told the judge. "I am no more guilty than

Gorbachev was for the crimes committed by his subordinates during the coup."

The judge laughed, but suspended the hearing. Brukhanov assumed that he had to telephone to Kiev for instructions. When the court reconvened that afternoon, his application was approved. He was a free man.

6

In Kiev Brukhanov's release went largely unnoticed; the minds of the Ukrainians were on other things. The failure of the coup had given them the chance to seize their independence, and under the cunning ex-Communist Leonid Kravchuk, the dream of centuries now came true. The role played by Chernobyl was given its due. Yuri Shcherbak, who described the coup as "a political Chernobyl," was made minister of the environment, and since its covert task had been accomplished, the membership of Green World dwindled from half a million to eighteen thousand. The nuclear power station at Chernobyl was not closed down. Despite a fire in the turbine hall of unit No. 3 the year after the coup, all three units were back on line by the beginning of 1993.

To the north, the professor of nuclear electronics, Stanislas Shushkievicz, who had had his instruments sealed away at the time of Chernobyl, became the first president of an independent Belorussia. In Moscow, Ligachev's rival, Alexander Yakovlev, took charge of Soviet radio and television; and Ilyn's antagonist, Professor Andrei Vorobyov, became Yeltsin's minister of health, while Ilyn, without the patronage of the once all-powerful Ministry of Medium Machine Building, saw his Institute of Biophysics disintegrate for want of funds.

Also in Moscow, in a grandiose flat with large rooms, tall ceilings and ornate fittings from the era of slaves and heroes, a small, stocky old man roamed among the memorabilia of his own heroic past. On walls hung ceremonial sabers commemorating the days when he had ridden with Budenny's cavalry brigade, the Konarmia, in the civil war. In a glass-fronted bookcase stood stacks of certificates in plump red folders extolling his achievements and recording his state awards. There were accolades from the Central Committee signed by the different general secretaries of the Communist Party of the Soviet Union, from Joseph Stalin to Mikhail Gorbachev.

This was Efim Slavsky, until Chernobyl the omnipotent minister of medium machine building. Cared for by his granddaughter, he was now almost stone deaf—but not so deaf that he had not heard and seen on television the demise of the Socialist state for which he had wielded his saber so many years before. What use now were those orders and accolades when the heroes of that era were called criminals by their former slaves? But the old man was still defiant. "They may destroy our Communist state," he shouted with clenched fist raised, "but they will never destroy the Communist spirit in me!"

Outside Moscow, in the wooded suburb of Zhukovska, ninety-two-year-old Academician Nikolai Dollezhal lived in retirement with his wife in the dacha given to him by Stalin for his role in the construction of the Soviet hydrogen bomb. Surrounded by tall pine trees, the brick house was now dilapidated, with ancient plumbing and peeling paint. In the living room there was a grand piano at which Academician Dollezhal had played until he had suffered a stroke a year or two before. Shostakovich had often played on this Steinway when he had dropped in from his home across the road.

In the garden was the small play house, its colors faded now in the summer sun. Here the childless Kurchatov had played with Dollezhal's daughter when he had spent the day with the Dollezhals. So too had other physicists who had received the same rewards for their services to Stalin's Soviet state—Academician Tam, Academician Khariton, Academician Kikoyin and even Academician Sakharov, until his second marriage, to Elena Bonner. There they met the artists and musicians who were also Dollezhal's friends, but never their fellow physicist, Academician Alexandrov.

Dollezhal had never liked politics; he had always refused to join the party, preferring a game of bridge to a discussion about Marx. Too much of his character had been formed before the Revolution; he had been raised as a Christian and still believed in God. Now, looking back on his life as old men do, he thought of how easily it might have taken a different course; he could have finished up like his school friend Salivsky in an outpost in the Sahara as a French legionnaire, or like his half-American friend, Vladimir Dixon, the son of the manager of the Singer Sewing Machine Company in Russia before the Revolution, who had volunteered to fight for the United States in World War I and had died in Paris of influenza.

Dollezhal had stayed behind, and in time had been swept up by the great Kurchatov, the mesmerizing "Beard," whose rages had frightened all those around him because it was known that he enjoyed Stalin's trust. Truly, Dollezhal had lived through a reign of terror, but he was a small man and it had not been hard to keep his head down. Now, in retirement, it was better to remember the good times and try to forget about the bad.

In another section of Moscow, a suburb in the time of Kurchatov but now absorbed into the city, the huge bronze head of the father of the Soviet atom bomb remained on its plinth outside the institute that he had founded. Behind it the thick steel gates, built to resist the forced entry of even the most determined saboteur, creaked as they opened on their rusty hinges.

Inside, the house formerly occupied by Kurchatov was now a museum, carefully preserving both the home and the archives of this Soviet hero. Visitors and interested scholars were told about the great man in hallowed tones. A television crew from Leningrad came to make a program about him, but to the consternation of the institute's scientists, a commentary was added with a sneering tone. "Yes," commented the reporter as the camera scanned the elegant living room with its comfortable sofas and chairs, "they really had a hard life, those pioneers of the Soviet state."

It was the same when journalists interviewed the honorary director of the Kurchatov, Academician Anatoli Alexandrov. He had retired now from running the institute, and his place had been taken by Academician Velikhov, Gorbachev's scientific adviser and Legasov's great rival, who was now vice president of the Academy of Sciences and had a fine office at the Neskuchny Palace. Alexandrov, however, retained all his titular honors, and on the rare occasions when he was driven in his huge Zil limousine to the institute he still sat in his huge office under a large portrait of Lenin to receive his guests.

"Anatoli Petrovich," asked a reporter from *Ogonyok* magazine, "a lot has been written about the Chernobyl disaster, but it would be nice to hear your opinion."

"You asked the question in a very delicate way," Alexandrov replied, "but in fact you are probably keen to find out whether I consider myself responsible for the accident. Don't feel awkward, I don't need it. . . . I tell you, Chernobyl became the tragedy of my life as well. I do feel it, every second. When it all happened, and I found out what they had done there, I nearly died. I was in a very poor state. Because of it, I decided to resign as president of the Academy of Sciences. I told Gorbachev about it. My colleagues tried to stop me, but I thought I had to do it . . ."

"Anatoli Petrovich, are there faults in the design of the Chernobyl-type reactor?"

"Yes, there are faults, but the cause of the accident was the defective experiment and gross violations of the regulations governing the operation of a nuclear power station. . . . The design cannot be blamed. Is the engine or the designer of a car to be blamed for an accident? Surely anyone can understand that it is the fault of the bad driver."

Whenever they could, the younger scientists at the Kurchatov tried to protect their chief from having to answer questions of this kind. Had not Professor Dubowski from Obninsk exonerated Alexandrov and stated that Dollezhal's bureau was to blame? Remembering the birthday party they had given for him in 1983 in the auditorium of the House of Culture, it took little imagination to realize how much the old man had suffered in the past eight years. His much-loved wife had died, his chosen successor, Legasov, had hanged himself, and there had been Chernobyl. Behind his haughty eagle eyes, and beneath that bald dome, there was surely a sorrow that no scientific formula could cure. Had not the finest minds of the greatest nation fought tirelessly to show the world that fascism and imperialism were not the only way? How many had died for socialism in Alexandrov's lifetime? Fifty million? Seventy million? How many more might have died if they had not armed themselves with an atom bomb? And what boundless opportunities this new source of power had presented. Had he not dreamed of nuclear heating for Arctic cities, nuclear ships, and nuclear trains? Had not his reactors embodied the promise of Soviet socialism?

Then came Chernobyl and the subsequent meltdown of the Soviet

Union. The first and only state built on the principles of science, with fuel rods assembled into a critical mass and control rods to prevent a runaway reaction. They had achieved the critical mass, and if people had behaved as predictably as atoms, it would have worked. But who could have foretold that all the control rods would be removed—and by the operators themselves? It was this that had led to the catastrophe: the human factor. But how could a physicist be expected to enter something so unpredictable into the equation? How hard it was that he had survived to see it all explode. In a generation in which so many lives had been too short, Alexandrov, like Dollezhal and Slavsky, was a man who had lived too long.

EPILOGUE

In January 1993 the International Atomic Energy Authority published a new report by its Nuclear Safety Advisory Group on the accident at Chernobyl, INSAG-7, which revised the analysis made in the first report in 1986, INSAG-1. The international experts now admitted the injustice of some of the charges made against the operators in INSAG-1—for example, that running the reactor at low power was forbidden by the regulations. This had been based "on oral statements made by Soviet experts during the week following the Vienna meeting. In fact, sustained operation of the reactor at the power level below 700MW(th) was not proscribed, either in design, in regulatory limitations or in operating instructions. The emphasis placed on this statement in INSAG-1 was not warranted."

INSAG-7 also revealed the findings of the Soviet commission headed by Nikolai Steinberg, published in Moscow in 1991, that the dangers inherent in the design of the control rods of the RBMK reactors had been noticed at the Ignalina nuclear power station in Lithuania in 1983. "Although the chief design engineer for RBMK reactors promulgated this information to other RBMK plants, and stated that design changes would be made to correct the problem, he made no such changes, and the procedural measures he recommended for inclusion in plant operating instructions were not adopted."

However, the authors of INSAG-7 did not entirely exonerate Dyatlov, Akimov and Toptunov. "INSAG remains of the view that in many respects the actions of the operators were unsatisfactory." The human factor was still considered a major element in causing the accident. The poor quality of operating procedures and instructions, and their conflicting character, may have put a heavy burden on the operating crew, and the type and amount of instrumentation, as well as the control room layout, made it difficult to detect unsafe reactor conditions. "However, operating rules were violated, and control and safety rods were placed in a configuration that would have compromised the emergency protection of the reactor even had the rod design not been faulty. . . . Most reprehensibly, unapproved changes in the test procedure were deliberately made on the spot, although the plant was known to be in a condition very different from that intended for the test."

INSAG-7 listed without comment the technical measures that had been taken in the former Soviet Union to improve the safety of the RBMK reactors, but conceded that the precise cause of the accident "may never be known." Indeed, the technical factors were less important than the political and psychological factors that had contributed to the accident, summarized as the absence of a "safety culture in nuclear matters, at national level as well as locally." It was left open to question whether or not this had been established in the now independent nations of Russia, Lithuania and the Ukraine.

The consequences of the accident remained as uncertain as the cause. In 1992 Western scientists confirmed reports from Belorussia of an increase in child thyroid cancer in the Gomel region. This had been expected, but not so soon. The overall incidence had risen from four cases a year between 1986 and 1989 to fifty-five in 1991 and sixty expected in 1992. In six cases, the cancer had spread, mostly to the lungs. One child had died and ten others were seriously ill.

Also predicted, but difficult to quantify, were the number of deaths from cancer which could eventually be ascribed to the collective dose of radiation received as a result of the accident. Robert Tilles, the director of Chernobyl Help in Moscow, thought it would ultimately claim more victims than World War II. The British National Radiological Protection Board estimated around 30,000 fatal cancers over the next 40 years in the affected parts of Russia and Western Europe, a 0.1% addition to those anticipated in the same population over the same period of time.

In a symposium on nuclear accidents in Helsinki in the summer of

1992, Abel Gonzalez, the Deputy Director of the Division of Nuclear Safety at the International Atomic Energy Agency in Vienna, accepted this "theoretical presumption . . . of several thousands of additional cancers," but felt it would never be susceptible to epidemiological detection because of the high background rate for cancer. In May 1992 a research project funded by the American Cancer Society, the British Imperial Cancer Research Fund and the World Health Organisation estimated that 250 million people alive today in the developed world would die prematurely from diseases caused by smoking.

INDEX

ABOUT THE AUTHOR

PIERS PAUL READ is the author of twelve novels
and two works of nonfiction. He studied history
at Cambridge University and in 1980 taught as
an adjunct professor of writing at Columbia University. He lives in London with his wife and four
children.

ABOUT THE TYPE

This book was set in Galliard, a typeface designed by Matthew Carter for the Mergenthaler
Linotype Company in 1978. Galliard is based on
the sixteenth-century typefaces of Robert Gran-
jon, which give it classic lines yet interject a
contemporary look.